Harold M. Pinsker

A CELL-BIOLOGICAL APPROACH TO LEARNING

Eric R. Kandel

Division of Neurobiology and Behavior
Departments of Physiology and Psychiatry
College of Physicians and Surgeons
Columbia University

Grass Lecture Monograph 1

Published by the Society for Neuroscience
Bethesda, Maryland

Library of Congress Catalog Card Number 78-69791
International Standard Book Number 0-916110-07-9

Based on the 1977 Grass Lecture at the
Seventh Annual Meeting of the Society for Neuroscience,
Anaheim, California.
This annual event is named for The Grass Foundation,
Quincy, Massachusetts, whose
endowment makes the lecture possible.

9650 Rockville Pike, Bethesda, MD 20014

Printed in the United States of America

PREFACE

This monograph is based on the Grass Lecture I had the privilege to present to the Society for Neuroscience on the evening of November 7, 1977. In keeping with the spirit of that evening, I have prepared the lecture for publication in the form that it was given, with only some stylistic alterations and slight elaborations of one point or another.

Much of the recent work reviewed in this lecture has been carried out in collaboration with either Vincent Castellucci, Thomas Carew, or Marc Klein. I am indebted to them and to several other colleagues in the Division of Neurobiology and Behavior at Columbia for their comments on an earlier draft of this lecture. I am also grateful to Sally Muir for help in preparing the manuscript for publication and to Kathrin Hilten for preparing the figures.

Throughout the period of research reviewed here, I have had a Career Scientist Award from the National Institute of Mental Health. In addition, my own research on learning and that of my colleagues has been supported by grants from the National Institute of Mental Health (26212), the National Institute of Neurological and Communicative Disorders and Stroke (12744), and the McKnight Foundation. I gratefully acknowledge this support.

Eric R. Kandel
Columbia University

April 1978

CONTENTS

INTRODUCTION

The most remarkable aspect of an animal's behavior is the ability to learn from experience, an ability that reaches its highest form in man. In good measure we are what we are because of what we have learned. In man, as well as in other animals, performance of most behaviors involves some aspects of learning and memory, and many psychological and emotional problems are thought to result, at least in part, from experience. Moreover, insofar as psychotherapeutic intervention is successful in treating mental disorders, it is presumably because treatment creates an experience that allows people to change.

The study of learning is central to psychology and neurobiology. Any meaningful mechanistic analysis of learning requires the combined use of the concepts and the methods of both disciplines. My purpose in this monograph is to outline some beginning steps toward an analysis of mechanism by reviewing recent attempts to combine behavioral and cell-biological approaches to the study of learning.

I should say at the outset that I use the term "learning" not in the restricted sense of some behaviorists who reserve it for only behavioral modifications, such as classical and instrumental conditioning, that are dependent on a specific association between stimuli or between a stimulus and a response (Hilgard and Marquis, 1940). Rather, I adhere to the broader sense of the term as used traditionally by ethologists and by many contemporary psychologists. In this broader sense learning refers to a modification of behavior resulting from the history of exposure to a stimulus, whether or not that stimulus is associated with other stimuli or responses. Memory, then, is the persistence of the learned modification—its storage, maintenance, and retrieval at various times after the completion of the learning task. According to this definition, learning includes a hierarchy of behavioral modifications ranging from simple non-associative forms—habituation, dishabituation, and sensitization—to more complex associative forms—classical and instrumental conditioning—to even more advanced forms such as the modifications produced by insight. This definition does not include, however, behavioral modifications that result from maturation, injury, or fatigue (see Schneirla, 1929; Humphrey, 1933; Harris, 1943; Thorpe, 1956; Harlow, 1958; Sokolov, 1963; Razran, 1971; Wagner, 1976).

I prefer the broad definition because it emphasizes that learning is not one, but rather a family of processes, presumably involving a variety of cellular mechanisms, each of which is worthy of study. In addition, the broader definition does not create artificial distinctions between behavioral modifications by placing exclusive importance on one aspect of learning—the presence or absence of associative specificity—to the exclusion of others, such as time course or adaptive significance.

My own interest in studying learning at the cellular level dates to 1958. This interest arose in part out of discussions with my friend and colleague Alden Spencer, and much of my subsequent thinking in this area has been shaped by my almost daily conversations with him. I am therefore saddened that he died, after a long illness, on Monday, November 1, 1977, one week before this lecture was given. Alden was a man of many talents and interests, but none matched his tireless enthusiasm for neural science and particularly his fascination with behavior. All of us who were touched by that enthusiasm have been greatly enriched.

It was in part as a result of Spencer's important early work (Spencer, Thompson, and Neilson, 1966*a,b,c*) that the study of learning, once considered beyond the reach of cellular methods, has begun to yield to these techniques. Now, a decade later, we are in a position to obtain a progress report and ask: What have we gained from this approach? What have we learned from applying cellular techniques to learning?

From the outset, the cellular approach held out the promise of a double benefit—methodological as well as conceptual. Methodologically, the cellular approach offered a set of powerful physiological, biochemical, and morphological tools for studying the neuronal mechanisms of behavior and learning—tools that had already proven immensely useful in the analysis of individual nerve cells. Conceptually, the cellular approach promised to serve a heuristic and clarifying function. The very attempt to translate behavior into cellular terms helps to clarify the study of learning. It forces us to examine the logic of various behavioral constructs, such as the distinctions between the different forms of learning and between the different phases of memory.

A significant contribution to the cellular study of learning has come from studies of invertebrates. I would like, in this review, to take stock of this contribution in a personal sense, by evaluating the utility to my work and that of my colleagues of the conceptual as well as the analytic aspects of the cellular approach. Let me begin with the conceptual aspects.

TASKS AND STRATEGIES IN THE CELLULAR STUDY OF LEARNING

The first conceptual contribution of the cellular approach was to specify clearly the requirements for effective study of the mechanisms of learning. In cellular terms these requirements could be formulated as a hierarchy of three interrelated ideas:

(1) defining, in complete cellular detail, the neural circuit of a behavior that is capable of being modified;
(2) locating in that neural circuit the critical neurons and interconnections that have been modified by learning; and
(3) analyzing the mechanisms of learning, first on the cellular and then (ideally) on the molecular level.

The critical feature of this outline is the emphasis on learning as a study with several interlocking levels, each of which is equally important. To study the mechanisms of learning, one must be able to analyze the wiring diagram of behavior. To analyze the wiring diagram of behavior, one must in turn be able to analyze the individual neurons that make up the wiring diagram.

This argument, which emphasizes the crucial role of analyzing behavior at the cellular level in elucidating the mechanisms of learning, is perhaps abstract. But it takes on further meaning if we consider two alternative strategies in the study of the neuronal mechanisms of learning.

At first thought, someone interested in the biology of learning and memory might be tempted to tackle immediately the problem in its most complex and interesting form—the higher forms of learning unique to the human brain, or at least those forms of learning that are common to man and other highly evolved mammals. In fact, this was the orientation that Alden Spencer and I originally had in 1958, when, at the start of our scientific careers, we joined forces at the National Institutes of Health to study the cellular properties of the hippocampus, that part of the mammalian brain thought to be most directly involved in some aspects of memory (Spencer and Kandel, 1961*a,b*; Kandel and Spencer, 1961; Kandel, Spencer, and Brinley, 1961). Despite the good progress we were able to make in the cellular analysis of these cortical neurons, it soon became clear to us that to understand how the cells of the hippocampus control behavior was formidable. The mammalian brain is immensely complex, both because of the number of neurons and because of the variety of their interconnections. As a result, it seemed unlikely that we would readily work out the wiring diagram of any behavior that could be modified by learning. Spencer and I therefore became convinced that if we were to bring the power of newly developed cellular techniques to bear on the study of behavior and learning, it would be necessary to develop manageable neural systems with greatly reduced numbers of neurons; an ideal neural system should contain at most a few thousand cells. Thus it seemed to us that a preparation could not be selected simply for its ability to learn; it had first to be selected for the ease with which its behavior could be analyzed at the cellular level.

Two types of systems have subsequently proven particularly useful in this regard: (1) simplified vertebrate systems such as the isolated spinal cord used by Spencer in his studies with Richard Thompson (Spencer et al., 1966*a, b*), and (2) higher invertebrates such as crayfish, insects, and snails (see for example Horridge, 1964; Kandel and Tauc, 1964; Krasne, 1969; Castellucci, Pinsker, Kupfermann, and Kandel, 1970; Callec, Guillet, Pichon, and Boistel, 1971; Zucker, Kennedy, and Selverston, 1971; Krasne and Bryan, 1973; Quinn, Harris, and Benzer, 1974; Gelperin, 1975; Woollacott and Hoyle, 1977; Kandel, Brunelli, Byrne, and Castellucci, 1976*a*; Davis and Gillette, 1978; Mpitsos, Collins, and McClellan, 1978).

I will here focus on the nervous systems of invertebrate animals. These nervous systems are attractive because of their numerical simplicity. The whole nervous system of an invertebrate animal generally contains only ten thousand to a hundred thousand nerve cells, a simplification that often allows one to relate the function of individual cells directly to behavior, the first key task in our hierarchy of tasks.

Although new in its application to behavior, a reductionist approach based upon the selection of technically congenial systems is traditional within biology. Advances have routinely depended on the selection and development of suitably simple experimental systems in which a family of interrelated problems can be studied effectively.

When it comes to behavior, however, one has somehow been reluctant to apply a reductionist's strategy. It is often said that behavior is the one area in biology in which simple animal models, particularly invertebrate ones, are least likely to be successful. The organization of the mammalian brain, and in particular the brain of man, is so complex that a comparative and reductionist approach to behavior based upon a study of invertebrates is bound to fail. Man has intellectual capabilities, language and abstract thinking, that are not found among simpler animals and that require types of neuronal organization qualitatively different from those found in invertebrates. Although these arguments are, in part, correct, they overlook certain critical issues. The issue is not whether there is something special about the human brain. There clearly is. The issue is rather whether the human brain and human behavior have *anything at all in common* with the brain and the behavior of simpler animals. If there are points of similarity, these might indicate that common principles of organization are involved, principles that could profitably be studied in simple animals.

The answer to the question of similarity is clear. Detailed work by students of comparative behavior such as Konrad Lorenz (1965), Niko Tinbergen (1951), and Karl von Frisch (1950) has shown that man shares many common behavioral patterns with simple animals, including elementary perception, motor coordination, and simple forms of learning. These findings, of course, do not guarantee that common neuronal mechanisms are actually involved in comparable aspects of behavior or learning in invertebrates and vertebrates, but the findings are encouraging and suggest that aspects of common mechanisms are likely to be involved. That the evolution of behavior is conservative should not be surprising. The evolution of other biological functions is similarly conservative. To choose a comparison that readily comes to mind, there are no fundamental functional or biochemical differences between the nerve cells and synapses of man and those of a snail. Since behavior is a reflection of nerve cell activity, it is perhaps not surprising that the behavior of man has at least some elementary features in common with that of the snail. For in the neuronal mechanisms of behavior, as in other areas of biology, similar solutions are likely to be found again and again throughout phylogeny. As a result, one is encouraged to think that a complete and rigorous analysis of behavior and learning in an invertebrate, no matter how simple the animal or the task, is likely to reveal mechanisms that will be of general importance. This rationale has provided the logical and philosophical basis for the current interest in the cellular mechanisms of behavior and learning in invertebrates.

APLYSIA: THE ADVANTAGES OFFERED BY INVERTEBRATE PREPARATIONS

Before considering a number of findings that have emerged from the cellular approach to learning, I would like to comment on the invertebrate animal that my colleagues and I have studied, the marine snail *Aplysia californica*. As is readily appreciated from its proud gait, illustrated in Figure 1, *Aplysia californica* is a large, attractive, and obviously intelligent beast. It is just the sort of animal one would immediately select for studies of learning.

Aplysia, in common with other invertebrates, has a nervous system that is numerically simple, containing about 20,000 central nerve cells (Figure 2). These cells are clustered into nine groups called ganglia, each containing about 2,000 cells. There are four paired ganglia (the cerebrals, buccals, pleurals, and pedals) and one unpaired ganglion, the abdominal. My attention first was attracted to *Aplysia*, and specifically to the abdominal ganglion, because three experimental features characteristic of higher invertebrates are particularly well developed in this ganglion (Figure 3).

(1) The nerve cell bodies are clustered around the periphery of the ganglion and can therefore be readily visualized.

(2) These nerve cells are unusually large; some of the cell bodies are among the largest somatic cells in the animal kingdom, reaching up to 1,000 μm in diameter.

(3) It is possible to record faithfully in the cell body of the neuron most synaptic and other subthreshold potentials generated on the axons and dendrites in the neuropil, to a degree that is generally not possible in the neurons of most other invertebrate animals. As in other invertebrate neurons, the synapses do not usually end directly on the cell body, but in *Aplysia* the synaptic region is electrically quite close to the cell body.

In the course of studying these cells, another important feature became apparent, a feature earlier appreciated by Arvanitaki and her collaborators (Arvanitaki and Tchou, 1942; Arvanitaki and Chalazonitis, 1958). In *Aplysia,* and in most invertebrates that have been studied, the same neuron can be identified repeatedly in different individuals and can be given a specific name (usually a prosaic one like R2, R15, L7).

The ability to identify cells repeatedly in every animal of the species makes it possible to map connections between cells on a single cell-to-cell basis (Figure 3*B*), and this in turn allows the examination of the pattern of interconnections. From the pattern of interconnections it is in principle only one small step to map connections between nerve cells and effector or receptor structures and thereby to work out the entire wiring diagram of a given behavior.

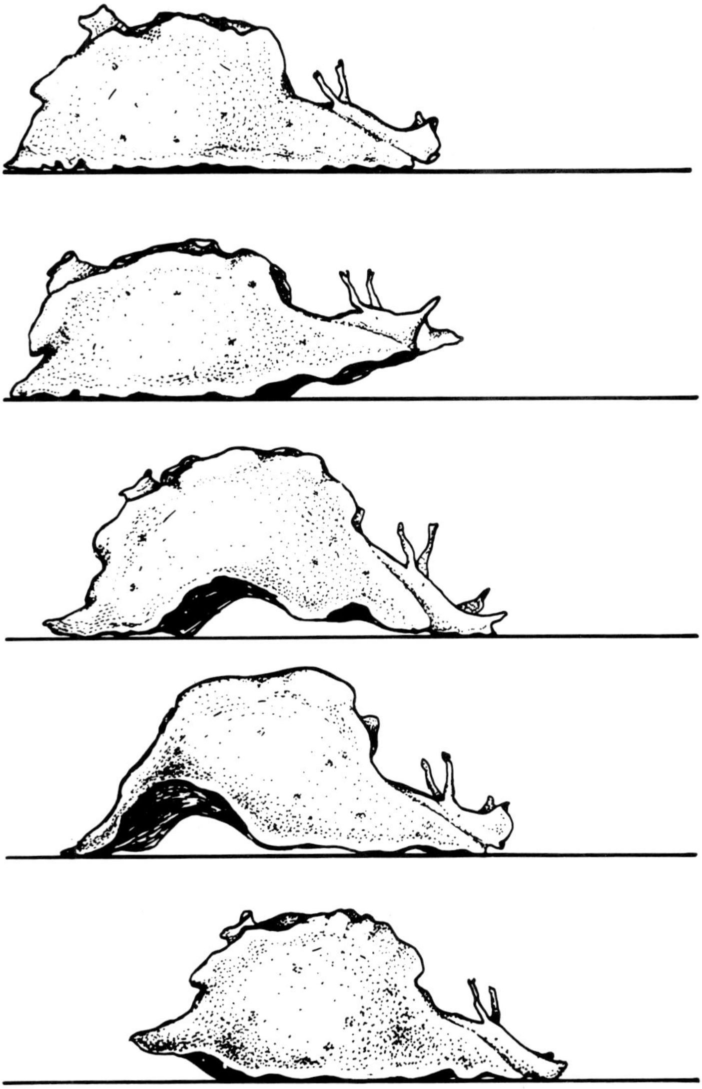

FIGURE 1. *Aplysia californica* crawling. This giant marine snail belongs to the opisthobranch subclass of the gastropod molluscs. Animals of this species may grow to be 30 cm in length and to weigh 1 kg.

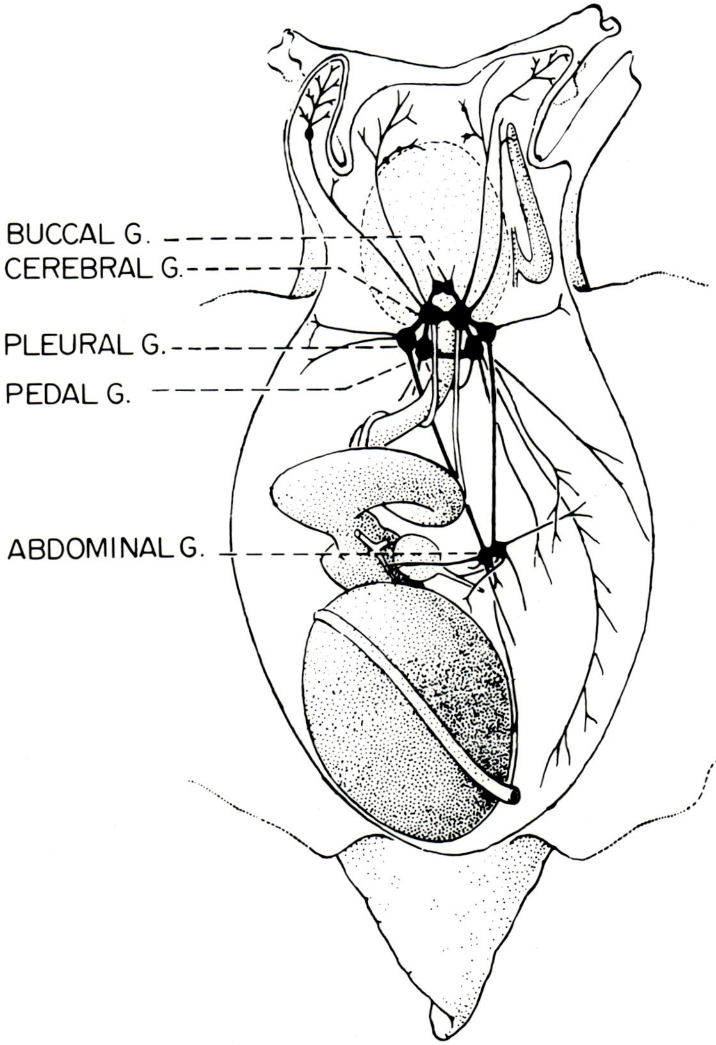

FIGURE 2. The nervous system of *Aplysia, in situ*, showing the four paired head ganglia and the single unpaired abdominal (or pleurovisceral) ganglion and their relation to other internal organs. (Modified from Eales, 1921.)

THREE IDEAS EMERGING FROM CELLULAR STUDIES OF INVERTEBRATES

The ability to study identified cells, their interconnections, and their relation to sensory and motor structures in *Aplysia* and in other invertebrates has in turn generated a number of findings that have caused us to look at the relationship between brain and behavior in a new way.

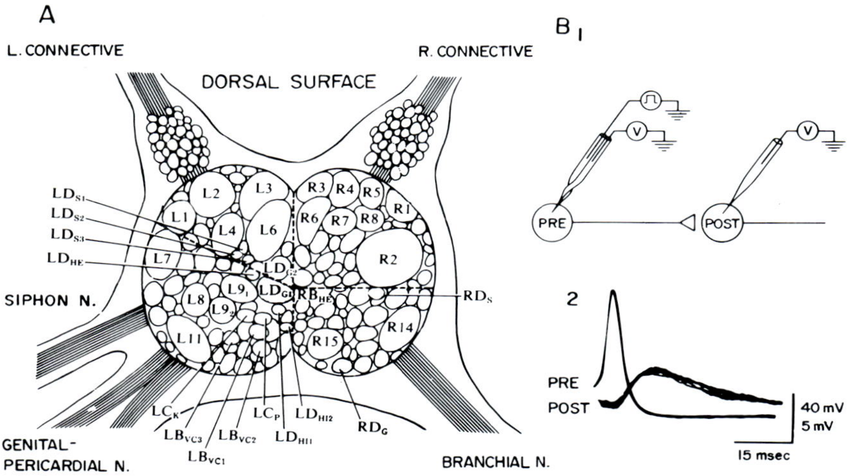

FIGURE 3. Invariance of cells and connections in the abdominal ganglion of *Aplysia californica.*

A: Map of identified cells in the abdominal ganglion, indicating their common position. The identified cells are labeled L or R (left or right hemiganglion) and assigned a number. Cells that are members of a cluster (consisting of cells with similar properties) are also labeled L or R. In addition, these cells are identified by cluster letter (LB, LC, LD) and a subscript identifying the behavioral function of the cell. Thus, cell LD_{HE} is a heart excitor belonging to the LD cluster, and cells LD_{S1}, LD_{S2}, and LD_{S3} are three different siphon motor neurons belonging to the same cluster. (From Frazier et al., 1967; Koester and Kandel, 1977.)

B: Mapping connections between identified cells. *B1* shows the experimental arrangement for simultaneous intracellular recordings from presynaptic and follower neurons. *B2*: A direct excitatory connection between two cells. The presynaptic neuron is depolarized and caused to fire repeatedly. Several oscilloscope sweeps are superimposed to illustrate the constant latency between the action potential in the presynaptic neuron and the excitatory synaptic potential in the postsynaptic cell. (From Kandel et al., 1967.)

The key findings that others working with various invertebrates and that my colleagues and I working with *Aplysia* have obtained can perhaps best be summarized in terms of three organizing ideas that have emerged from this work.

(1) *Many neurons of a brain are unique individuals that make highly specific connections with other neurons or effector organs.*

The first part of this generalization, that the various nerve cells of an invertebrate brain are unique individuals, was appreciated initially by Richard Goldschmidt (1907) in his studies of the parasitic worm *Ascaris*. His idea has now been well documented for many higher invertebrates that have been studied on the cellular level. Using modern techniques we can now examine the factors that account for the differences between neurons. For example, in the abdominal ganglion of *Aplysia*, neighboring cells may have characteristically different firing patterns. One cell may be silent while its neighbor may be spontaneously active. Among cells that are active, some cells fire regular impulses (action potentials) whereas others fire impulses in recurrent brief bursts or trains (Frazier, Kandel, Kupfermann, Waziri, and Coggeshall, 1967; Strumwasser, 1967). These characteristic firing patterns arise because the excitable membranes of the cell bodies of neurons are complex and capable of producing up to six or seven ionic currents in varying combinations (Connor and Stevens, 1971*a,b*; Wilson and Wachtel, 1974; Meech and Standen, 1975; Smith, Barker, and Gainer, 1975; Eckert and Lux, 1976; Thompson and Smith, 1976; for review see Kandel, 1976). This contrasts with the membrane of most axons—such as that of the squid—which produces only two currents when active. In addition to variations in firing patterns, neurons also vary in size, position, shape, connections, and transmitter biochemistry (see for example Otsuka, Kravitz, and Potter, 1967; Giller and Schwartz, 1968; McCaman and Dewhurst, 1970; Eisenstadt, Goldman, Kandel, Koike, Koester, and Schwartz, 1973). Because of these findings it is beginning to seem probable that many—although not all—neurons are unique individuals. Whether or not most cells in the mammalian nervous system are also distinct individuals is not yet known. However, the studies in the visual systems of mammals by Hubel and Wiesel (reviewed in the Society for Neuroscience 1976 Grass Lecture) and those of Mountcastle and his colleagues on the somatosensory and parietal cortex (reviewed in the Society's 1976 Special Lecture) have revealed fascinating and important differences between neighboring neurons.

The earlier studies of the development of the vertebrate brain by Roger Sperry led him to a similar conclusion (Sperry, 1951, 1963).

In addition to neurons being unique individuals, the connections between neurons are also precise and invariant. Studies in *Aplysia* and in other invertebrates have shown that a specific neuron makes the same kinds of connections to other cells in every member of the species. This applies not only to the *presence* of the particular connections; it also holds for the *sign* of the connections—whether the connections are inhibitory or excitatory. In fact, individual cells may mediate different actions to different cells, and the sign of these different actions is invariant. For example, a double-action cell will always make excitatory connections to one type of cell and inhibitory connections to another type (Kandel, Frazier, Waziri, and Coggeshall, 1967; Wachtel and Kandel, 1967; Kehoe, 1969, 1972; Gardner and Kandel, 1972; Gerschenfeld and Paupardin-Tritsch, 1974; Berry and Cottrell, 1975; Kandel, 1976; Weinreich, 1977; McCaman and McKenna, 1978).

(2) *Individual cells have specific and often unique roles in behavior.*

In the course of studying how nerve cells control behavior, a nice surprise was encountered: Individual cells proved to exert not only a specific but sometimes a powerful control over behavior in a variety of invertebrates (Wiersma, 1952; Willows, 1967; Kupfermann and Kandel, 1969; Stuart, 1969; Koester, Mayeri, Liebeswar, and Kandel, 1974; Carew and Kandel, 1977).

Moreover, different classes of behavior seem to utilize neurons with different biophysical properties and patterns of interconnections. Thus one may ultimately be able to develop rules that relate the various classes of behavior (such as graded reflex responses on the one hand and all-or-none fixed acts on the other) to various types of neural circuits (Kandel, Carew, and Koester, 1976*b*).

One final behavioral feature of certain invertebrates, such as the opisthobranch molluscs, is the animal's rather limited behavioral repertory. This feature, which at first seems a disadvantage, ultimately can be used advantageously. For example *Aplysia* generates only a few main types of behavior (Figure 4). It has therefore been possible over the last 10 years for my colleagues and me, and for Irving Kupfermann and his colleagues, to work out a significant portion of the animal's total behavioral repertory on the cellular level. We have now worked out representative aspects of locomotion, defensive and escape responses, reproductive behavior, feeding, and the neural control of the osmotic,

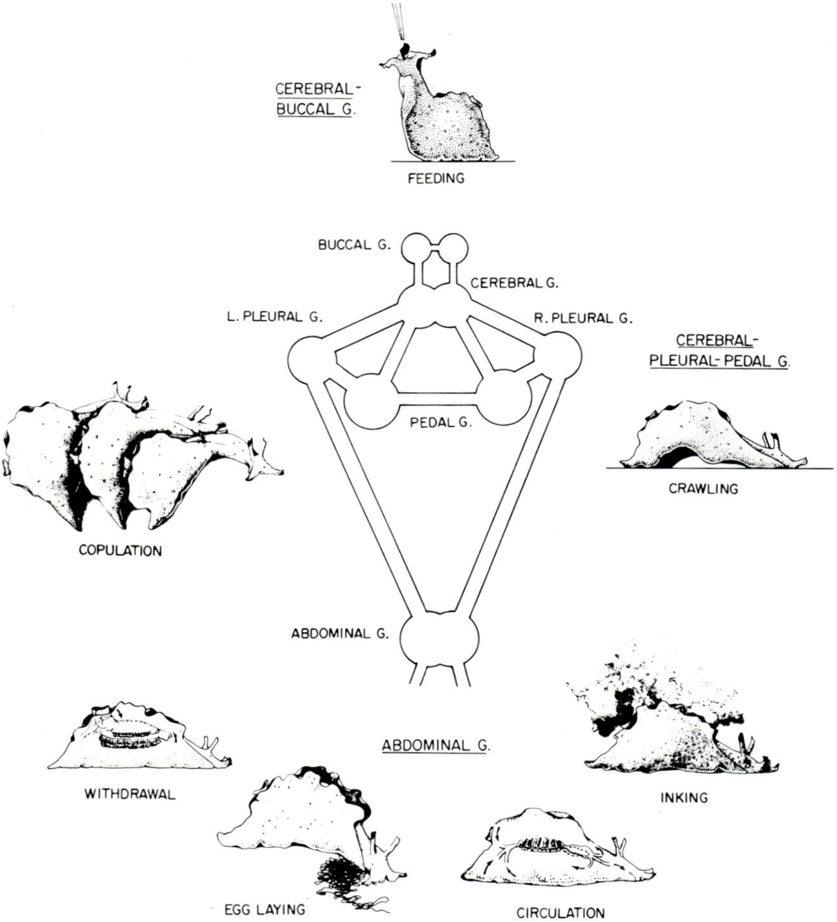

FIGURE 4. Behavioral repertory of *Aplysia*. Many of the key behaviors of the animal's repertory have been analyzed, at least in part, on the cellular level. The drawing for each behavior is placed near the ganglion that contains the critical motor cells underlying the behavior.

respiratory, and circulatory adjustments that accompany some of these behaviors. My colleagues and I started at the rear end of the animal and focused on the behavior controlled by the cells that we had earlier identified in the abdominal ganglion. These cells control defensive behaviors such as withdrawal and inking, reproductive behaviors such as egg laying, and a variety of respiratory and circulatory adjustments (Kupfermann and Kandel, 1969, 1970; Carew, Pinsker, Rubinson, and Kandel, 1974; Koester et al., 1974; Kupfermann, Carew, and Kandel, 1974; Mayeri, Koester, Kupfermann, Liebeswar, and Kandel, 1974; Carew and Kandel, 1977; Dieringer, Koester, and Weiss, 1978). Recently Hening, Carew, and I have made an excursion into the pleural and pedal ganglia to explore the neural mechanisms of locomotion, including escape galloping (Hening, Carew, and Kandel, 1977). Starting at the head of the animal, Kupfermann and his colleagues have focused on the buccal and cerebral ganglia in the course of analyzing feeding, a complex fixed-action pattern (for review see Weiss and Kupfermann, 1978). Ultimately we hope to join forces in putting together aspects of the animal's whole behavioral repertory in terms of the various neural circuits that mediate it. This type of effort will be necessary for analyzing the mechanisms of arousal and behavioral choice, processes that cut across simple behavioral categories and involve interactions among several behavioral systems. I shall, at the end of this lecture, return briefly to this theme in considering cellular studies of arousal. (For other discussions of these problems see Davis, Mpitsos, and Pinneo, 1974; Advokat, 1978; Hawkins and Advokat, 1978.)

(3) *Certain types of learning result from changes in the activity of specific cells and from changes in the connections between cells.*

The finding that behavior is mediated by invariant cells that interconnect in precise ways made it practical to combine behavioral and cellular methods to examine in detail the locus and the mechanism of a particular behavioral modification. These studies indicate that certain types of learning can be localized to specific sites in the nervous system and that changes produced by learning involve alterations in the effectiveness of previously existing synaptic connections. This is the topic I will consider in the main body of this lecture.

By using a behavioral system in which a variety of types of learning can be produced, it is possible not only to explore the locus and mechanisms of each type of learning, but in addition to begin to examine how the loci and mechanisms of one type relate to those of another. This allows an examination of a number of key questions in the current study of learning which concern the interrelationships between different types of behavioral modifications. For example, one could examine how a simple form of non-associative learning like habituation relates to a more complex form like sensitization, how short-term memory relates to long-term memory, and, in the long run, how non-associative learning relates to associative learning and how certain abnormalities in behavior are learned.

Here I would like to review three aspects of our work. First I will describe the system that we have studied and review the capabilities for learning that this system exhibits. Second, I will consider studies on the cellular mechanisms of habituation, a simple form of learning that we have used to explore the mechanistic interrelationship between short- and long-term memory. In particular, I will describe recent experiments designed to investigate one of the central questions in the study of learning—whether the memory for short-term and long-term habituation involves two separate and distinct memory traces with different loci and different mechanisms or whether both involve a single memory trace that changes in character. Third, I will discuss studies of sensitization, a second and slightly more complex type of learning, and describe our attempts to relate biochemical and electrical changes in the presynaptic terminals to the synaptic plasticity underlying short-term behavioral modification. At several points I will emphasize the limitations in our current understanding and indicate experiments needed to move beyond them.

HABITUATION: AN ELEMENTARY FORM OF LEARNING

The behavior that we have used to study these questions is one of the simplest the animal has, a defensive withdrawal reflex of the external organs of the mantle cavity.

The mantle cavity of *Aplysia* is a respiratory chamber that houses the respiratory organ, the gill. This chamber is covered by a protective sheet, the mantle shelf, which terminates in a fleshy spout, the siphon. When a weak or moderately intense tactile stimulus is applied to either the siphon or the mantle shelf, both the siphon and the gill contract and withdraw into the mantle cavity (Figure 5*A*).

The two components of the reflex—the withdrawal of the gill and of the siphon—parallel each other, so that observing one component provides an index of the other, and both components are concomitantly modified by appropriate training procedures (Carew, Pinsker, and Kandel, 1972; Pinsker, Hening, Carew, and Kandel, 1973). Because the siphon normally protrudes from the mantle cavity, this component of

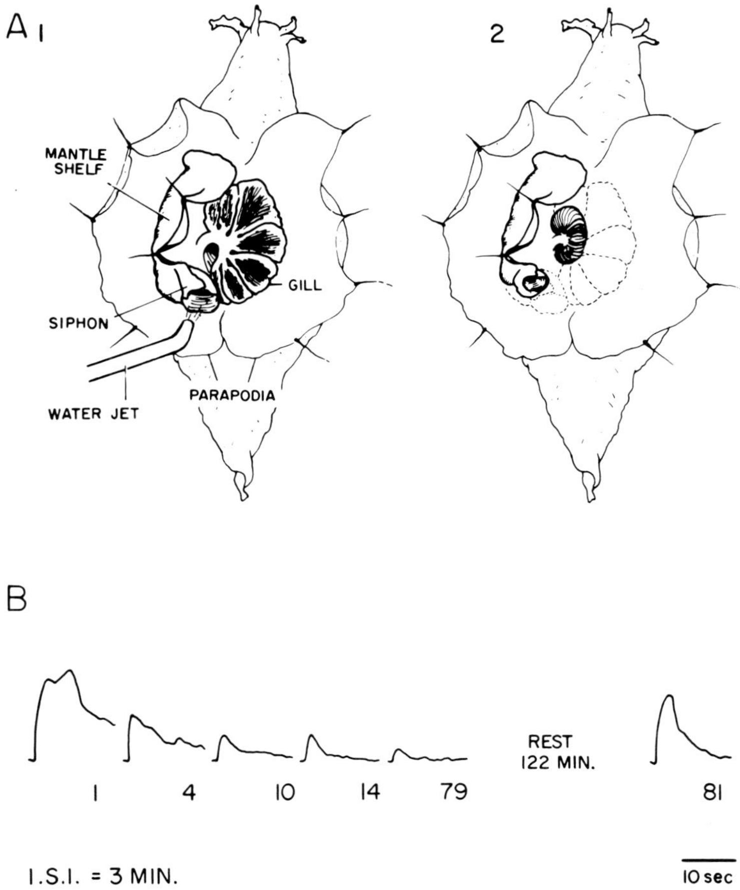

FIGURE 5. Habituation of defensive withdrawal.

A: The defensive withdrawal reflex of the siphon and gill in *Aplysia* (dorsal view of an intact animal). Since the parapodia and mantle shelf ordinarily obscure the view of the gill, they must be retracted to allow direct observation. The tactile receptive field for the gill-withdrawal reflex consists of the siphon and the edge of the mantle shelf. *A1:* Relaxed position. *A2:* Defensive withdrawal reflex in response to a weak tactile stimulus to the siphon (a jet of seawater delivering an effective pressure of 250 g/cm^2). The relaxed position of the gill is indicated by the dotted lines.

B: Habituation and spontaneous recovery of the gill-withdrawal reflex. Habituation of the response with 80 repetitions of the stimulus at 3-min intervals. The major decrease occurs within the first 10 stimuli. Following a 122-min rest, the response recovered partially. (From Pinsker et al., 1970.)

the reflex can be studied readily in freely moving animals and is particularly useful for studying long-term behavioral modifications. The gill is normally covered in part by the mantle shelf, and is best studied in a restrained preparation where the mantle shelf can be retracted. This component is therefore most useful for studying short-term behavioral modifications or for examining, in restrained animals, the consequence of long-term siphon training. As we will see later, the advantage of the gill component is that its neural circuit is somewhat simpler and better studied than that of the siphon component.

This withdrawal reflex is analogous to defensive withdrawal responses found in almost all higher invertebrates as well as in vertebrates. A significant feature of this and other defensive reflexes is that they can readily undergo two *elementary forms* of learning, habituation and sensitization. I will first focus on *habituation.*

Habituation is probably the most ubiquitous behavioral modification found in animals, including man. It refers to a decrease in behavioral response that occurs when an initially novel stimulus is repeatedly presented. Habituation is the means whereby animals and man learn to ignore stimuli that have lost novelty or meaning. Habituation not only is readily achieved in the laboratory but abounds in nature. For example, if a novel stimulus, such as a sudden noise, is presented for the first time, attention is immediately drawn to it and heart rate and respiration rate may change. However, if the same noise is repeated, attention and bodily responses gradually diminish so that one may become unaware of the noise unless aroused (dishabituated) by some other stimulus. One also becomes habituated to some bodily sensations, which enter awareness rarely and only under special circumstances. In this sense, habituation is learning not to respond to trivial recurrent external or internal stimuli. The elimination of response patterns that fail to serve useful functions is as important for survival as the development of new responses. This fact can perhaps best be seen in newborn animals, who often make escape responses to a variety of nonspecific stimuli. Most of these cease to be effective with time, but the effectiveness of some persists. As a result of the habituation of many responses, the remaining responses become focused on stimuli that persist in novelty or that become associated with satisfying or alarming consequences.

The importance of habituation in the development of selective perception and responsiveness is illustrated in the escape response of birds to a flying hawk. Using models of a hawk, Tinbergen (1951) found that fleeing occurred when the hawk flew overhead in the forward, but not the backward, direction. Schleidt (1961*a,b*) reexamined this response using turkeys raised in a controlled environment. He found that when the birds were first exposed to circular or square models overhead they responded by fleeing. With repeated exposure they habituated to these stimuli, but they still gave fleeing responses when occasionally exposed to a hawk model. However, when turkeys were repeatedly exposed to hawk models they habituated to these and fled in response to circles or squares. Turkeys therefore show selective escape responses to hawks because they become habituated to the other birds who frequently fly overhead, but not to hawks, which are seen rarely. Thus habituation serves in a variety of contexts as a learned adaptational mechanism designed to eliminate responses to biologically irrelevant stimuli and to restrict maximal responsiveness to situations that enhance survival.

Habituation has been found in a variety of behavioral response systems in all animals, including man. Simple reflexes, such as the abdominal flexion reflexes in humans, show habituation, as do complex reflexes, such as the orienting response.

The general occurrence of habituation, its adaptational importance, as well as its simplicity, have made it an attractive paradigm for studying the neuronal mechanisms underlying an elementary form of learning. In fact, it is the widespread occurrence of habituation and the substantial progress that Spencer and Thompson (Spencer et al., 1966*a,b,c*) had made in analyzing habituation of the flexion reflex in the spinal cord of the cat that first focused my attention on this form of learning (Kandel and Tauc, 1965).

An interesting aspect of habituation, illustrated by the siphon- and gill-withdrawal reflex in *Aplysia*, is that it can have both a short- and a long-term form (Pinsker, Kupfermann, Castellucci, and Kandel, 1970; Carew et al., 1972). Following a single training session of 10 to 100 stimuli spaced 30 seconds to 5 minutes apart, full recovery generally occurs within a day and partial recovery can occur within minutes or hours (Figure 5*B*). However, with 4 training sessions of 10 stimuli each, repeated on consecutive days, retention is prolonged and lasts days and even weeks (Figure 6). Recovery, or its inverse, retention, is a measure of memory in this type of learning; it is equivalent to forgetting in other types of learning. Thus habituation is a simple type of learning with both a short- and long-term memory, a feature that allows the examination of the relationship between the two.

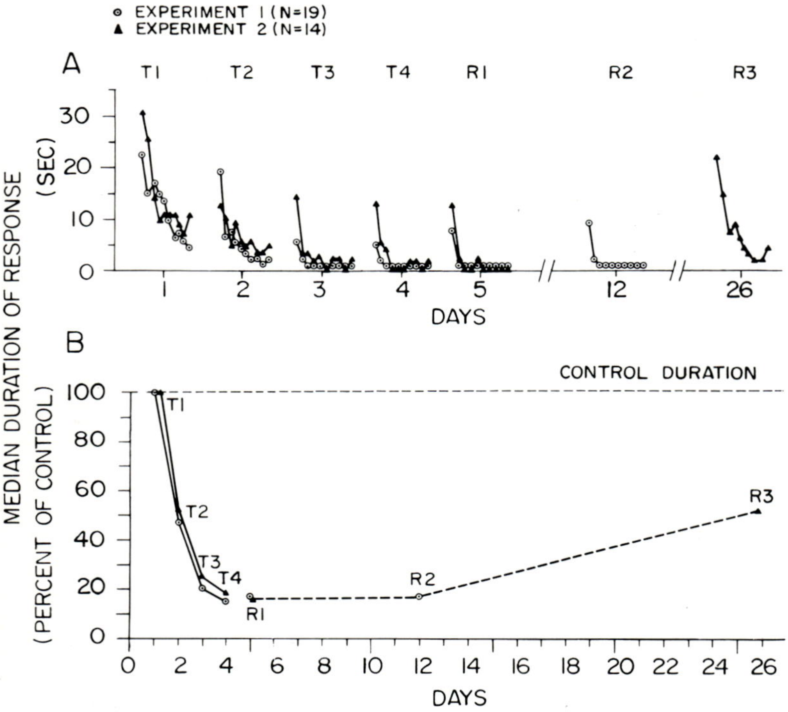

FIGURE 6. Long-term habituation of siphon withdrawal in *Aplysia.*

A: Habituation during 4 daily sessions of training (T1 to T4), and retention 1 day (R1), 1 week (R2), and 3 weeks (R3) after training. Each session consisted of 10 trials with an interstimulus interval of 30 sec (each trial is a single stimulus). Data from two experiments are compared. In experiment 1, retention was tested 1 day (R1) and 1 week (R2) after training. In experiment 2, retention was tested 1 day and 3 weeks (R3) after training. Each data point is the median response of the population in a single trial.

B: Time course of habituation, based on the experiments illustrated in part *A*. The score for each daily session is the median of the sum of 10 trials given to each animal. Control duration (100%) is the response time during the first day of training. T1 to T4 indicate the scores for 4 daily training sessions. Retention was tested 1 day (R1), 1 week (R2), and 3 weeks (R3) after 4 days of habituation training. (From Carew et al., 1972.)

Wiring Diagram of the Gill-Withdrawal Reflex

We next worked out what we believe to be a nearly complete wiring diagram for both the gill and the siphon component of the reflex as elicited by weak or moderate intensity stimuli to either the mantle shelf or the siphon (Figure 7; see also Kupfermann and Kandel, 1969; Castellucci et al., 1970; Kupfermann et al., 1974; Byrne, Castellucci, and Kandel, 1974, 1978). I will focus here only on the gill component as elicited by siphon stimulation, because it is better understood (Figure 8). The neural circuit for this component is remarkably simple and illustrates clearly the advantages of using an invertebrate preparation for working out the neural circuit of a behavior. There are 6 motor cells to the gill; these cells receive information from the siphon skin by means of about 24 sensory neurons (of which only 8 are illustrated in Figure 8) and several interneurons, 3 of which—2 excitatory and 1 inhibitory—are illustrated here. The motor cells receive information from the skin of the mantle shelf from a separate cluster of sensory neurons, as illustrated in Figure 7. Stimulating the siphon skin activates the sensory neurons that innervate the siphon; these then excite the interneurons and the motor cells by means of direct monosynaptic connections. The motor cells in turn connect directly to muscle. There are two striking features about this neural circuit: (1) its numerical simplicity, and (2) the repeating pattern of connections: each gill motor neuron receives qualitatively similar inputs from each of several sensory neurons innervating the siphon.

I will first discuss the numerical simplicity. Even though we are certainly missing several interneurons, the circuit probably does not involve more than 35 to 40 cells. As a result, the relation of specific sensory or motor cells to behavior can be determined in a causal and quantitative way. For example, because there are only six motor cells, we can readily examine the contribution of each to the behavior. Thus, by hyperpolarizing any given motor cell we could reversibly remove its contribution to the behavior and determine how much the behavior is reduced. Alternatively, we can directly stimulate the motor cell in a way that simulates the firing pattern produced by the natural sensory stimulus and determine whether the cell's ability to generate behavior matches quantitatively the contribution that is removed when the cell is hyperpolarized (Figure 9*A*). Using reversible blockage and intracellular stimulation techniques, we have found that three of the cells are major motor cells (L7, LD_{G1}, LD_{G2}) and collectively contribute about 70%

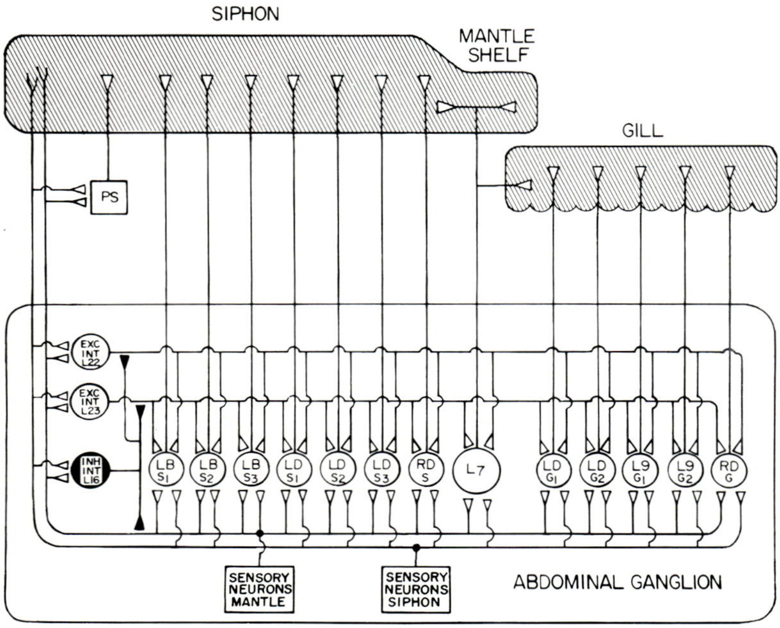

FIGURE 7. Schematic drawing of the neural circuit that mediates the defensive reflex withdrawal of the gill and siphon to weak or moderate intensity stimuli, indicating the sensory, interneuronal, and motor-neuronal components of the total reflex pathway. One population of sensory neurons innervates the siphon skin; another innervates the mantle shelf. Each consists of about 24 cells. These neurons have direct connections to the motor neurons and indirect connections via two excitatory interneurons and one inhibitory interneuron. The 3 interneurons and the 13 central motor cells in the abdominal ganglion are all unique, identified cells. The peripheral siphon motor cells (PS) are not uniquely identified. The two populations of sensory neurons activate both the siphon and gill motor neurons; this accounts for the fact that the two acts of the reflex pattern occur simultaneously. In this figure and all subsequent figures, black triangles represent inhibitory synapses, white triangles represent excitatory ones. (From Kupfermann et al., 1974; Bailey et al., 1978.)

to 80% of the total gill-withdrawal reflex, whereas the remaining cells are minor motor cells that contribute much less (Kupfermann, Pinsker, Castellucci, and Kandel, 1971; Kupfermann et al., 1974).

We also examined the contribution of a sensory cell to the total behavior (Figure 9*B*). Although there are 24 cells, the sensory cells have overlapping receptive fields, so that stimulating a particular point on the

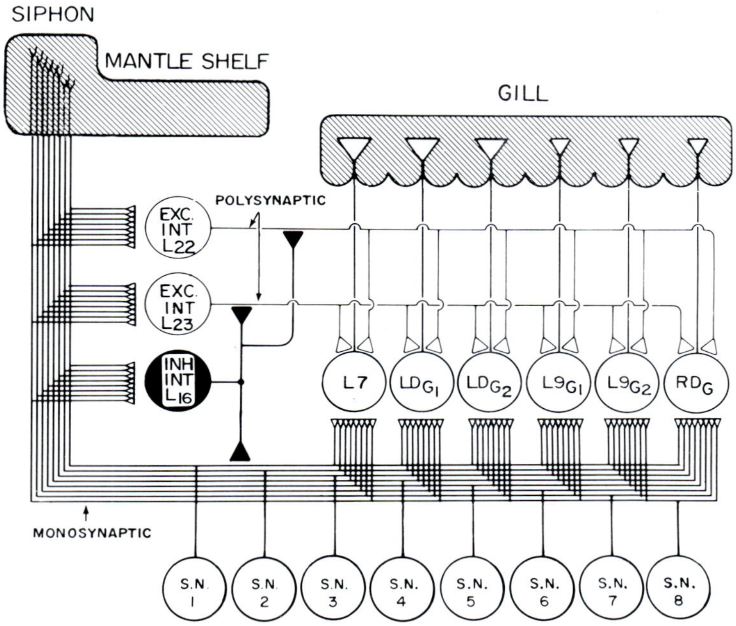

FIGURE 8. Simplified wiring diagram of the defensive gill-withdrawal reflex. The siphon skin is innervated by approximately 24 mechanoreceptor sensory neurons. The diagram has been simplified to illustrate that a tactile stimulus to a point on the skin only excites approximately eight sensory neurons. The sensory neurons make direct (monosynaptic) connections to six identified gill motor neurons and to at least one inhibitory (L16) and two excitatory (L22, L23) interneurons. (From Kandel et al., 1976*a*.)

siphon skin with a small probe activates only about 8 cells. Since a weak stimulus tends to produce only one to two spikes in each sensory neuron, the system can be reduced even further, and a single sensory cell can be fired repetitively to elicit the behavior. With these types of experiments, which are essential for a causal study of behavior, one can show that activity in each member of a group of cells is both necessary and sufficient for a component of the behavior.

Let me now turn to the related issue of the repeating patterns of connections. As is evident in the circuit shown in Figure 8, the behavior is mediated by a number of modules, or repeating units, in parallel. Each piece of skin is innervated by a different subpopulation of similar sensory neurons, and these connect to motor cells, each of which in-

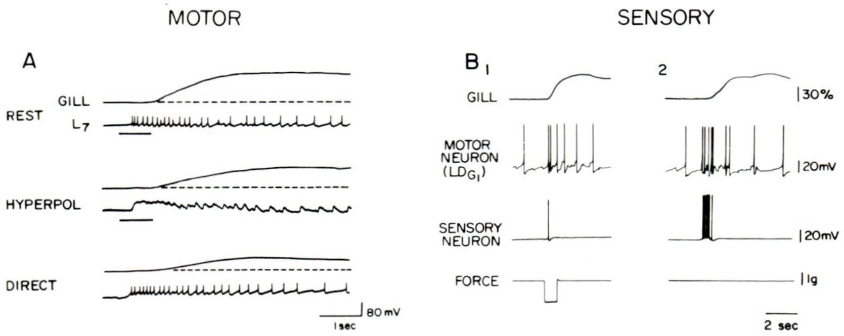

FIGURE 9. Contribution of individual motor and sensory cells to the gill-withdrawal reflex.

A: Motor cell. Effects on the gill-withdrawal reflex of functional removal from the neural circuit of the behavior of gill motor cell L7. The gill-withdrawal reflex was elicited by a jet of seawater (indicated by the short bar) applied to the siphon every 5 min. On alternate trials cell L7 was hyperpolarized so that the excitatory input could not discharge it. In the last set of traces L7 was directly fired by a long depolarizing pulse that was adjusted to fire L7 in a pattern comparable to that seen during the gill reflex. The size of the gill contraction due to direct firing of L7 was approximately equal to the reduction of gill contraction caused by removing L7 from the reflex. (Modified from Kupfermann et al., 1974.)

B: Sensory cell. *B1*: Stimulation of a point on the skin produces a discharge of the monitored sensory cell and repetitive firing of the monitored motor cell. The gill contraction and the applied force were recorded simultaneously. *B2*: Since each point on the skin activates about eight sensory cells, each of which fires one to two spikes, one can attempt to simulate the natural behavior of the total population of sensory cells by firing a single cell repetitively, producing eight to ten spikes. This procedure mimics the natural input of the entire population of sensory cells to the motor cells, causing them to fire and to produce gill contractions similar to those produced with natural stimulation. (From Byrne et al., 1978.)

nervates a different combination of gill muscle fibers. Therefore, as stimulating a single sensory neuron has indicated (Figure 9*B*), we can restrict the behavior to a single module, a microcircuit consisting of only a single sensory neuron, a major motor neuron, and the several intervening interneurons. In fact, by looking at a single sensory neuron and a major motor neuron we can obtain a realistic view of the behavior from the perspective of *only two cells* and the synapse between them. This is not only a model of the total behavior, but a significant component of the whole. Much of what I will consider later will be based on an examination of the behavior from this vantage point.

Analysis of Short-Term Habituation

Having worked out the neural circuit of the behavior, we were now in a position to analyze it during behavioral habituation. First we focused on the short-term process. Here we wanted to distinguish between two alternative mechanisms for short-term memory: (1) a dynamic, reverberatory change—a circulation of impulses in a closed neuronal chain—of the sort proposed by Hebb (1949), leading perhaps to a build-up of inhibition, or (2) a plastic change in synaptic efficacy, first proposed by Ramón y Cajal (1909) and later elaborated by Konorski (1948). If the change proved plastic, we also wanted to specify its *locus*. In particular, we wanted to know if the plastic change is restricted to one set of neurons or distributed among many.

By examining the various sites within the circuit during habituation both in the intact animal and in various simplified preparations (Figure 10), Castellucci, Kupfermann, Pinsker, and I found that the change underlying short-term habituation was a plastic change in synaptic effectiveness—a depression in excitatory transmission that was specifically localized to the synapses made by the sensory neurons on their central target cells, the motor neurons and the interneurons (Figure 11, and Castellucci et al., 1970; Kupfermann and Kandel, 1970; Kupfermann et al., 1971). This type of localization proved fortunate, for it allowed us to examine what happens during habituation by analyzing changes in monosynaptic connections; that is, we had isolated a neural correlate of a short-term form of memory in a monosynaptic system.

These monosynaptic connections can be examined by recording the unitary excitatory postsynaptic potentials (EPSPs) produced in the motor cells by the stimulation of a single sensory neuron (Figure 12). During a single training session of 10 to 15 stimuli, repeated firing of the sensory neurons at rates that produce habituation causes the EPSP to undergo a depression that parallels the habituation. A second training session produces a build-up of depression. With further training sessions the EPSP can be completely depressed to the point where it is no longer detectable.

The question now arises: How are these changes produced? Are they presynaptic or postsynaptic? This question can be answered by means of a quantal analysis. As reflected in the graded nature of the EPSP, transmitter release appears to be graded, but it is actually quantized. Each quantum contains several thousand molecules of transmitter that are released in an all-or-none way to produce by their action

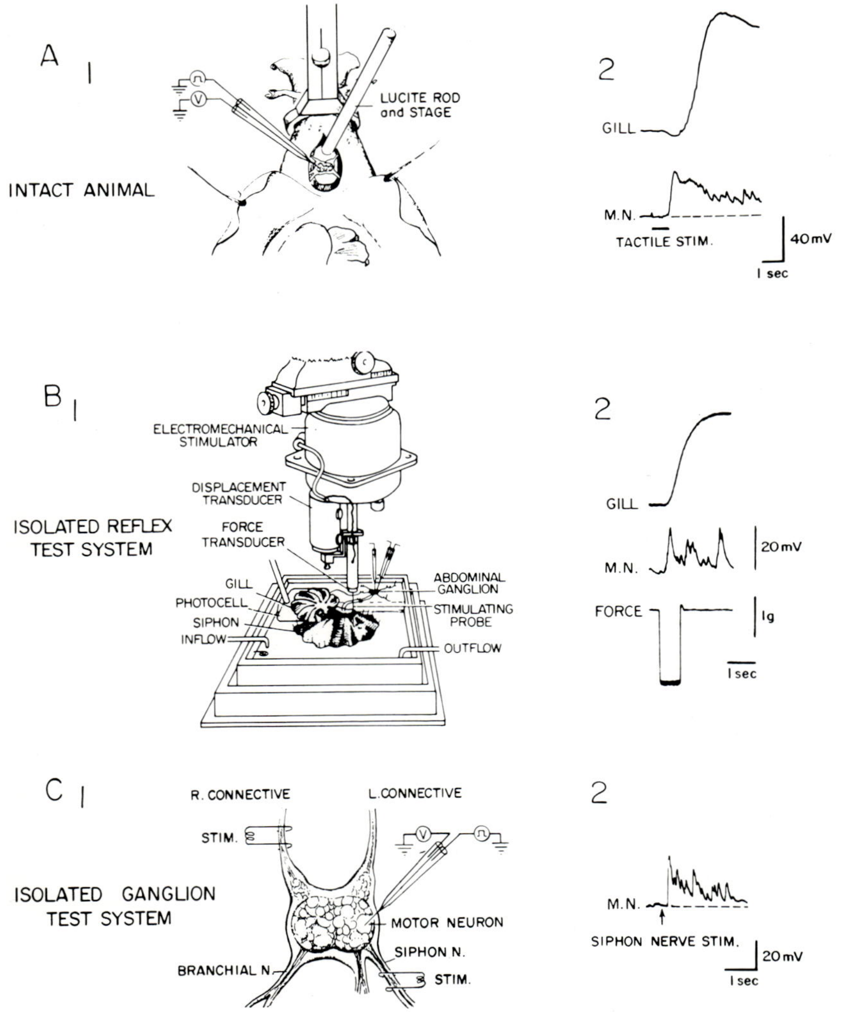

FIGURE 10. Comparison of recordings from an intact preparation with a slit in the neck and those from two simpler test systems.

A: Intact preparation. *A1*: Experimental arrangement. Intracelluar recordings are obtained from major gill motor cell L7. *A2*: Simultaneous recordings of gill contraction (obtained with a photocell) and a complex EPSP in L7 in response to an 800-msec jet of seawater applied to the siphon. Cell L7 was hyperpolarized about 50 mV to prevent spike generation. The remaining gill contraction is due to the activity of the other gill motor neurons in the reflex.

a small, all-or-none, unit synaptic potential (del Castillo and Katz, 1954*a,b,c*; Hartzell, Kuffler, and Yoshikami, 1976).

A synaptic potential is typically made up of many unit potentials, because each impulse typically releases many quanta. The average amplitude of a synaptic potential, $\bar{E}$ (the synaptic efficacy), produced by a presynaptic impulse is given by:

$$\bar{E} = m \times \bar{q}$$

where $\bar{q}$ is the average amplitude of the unit synaptic potential or the potential produced by a single quantum, and m is the average quantal content or the average number of quanta released in a series of stimuli. Thus the EPSP is generated by an integral number of quanta, m, each of which produces a potential change of size, $\bar{q}$, in the postsynaptic cell.

Analysis of synaptic transmission in terms of its quantal components therefore provides a means for using the synaptic potential to assay electrophysiologically changes in the size of individual quanta or in the number of quanta released by an action potential in the presynaptic terminals.

A quantal analysis of synaptic action is powerful because it can reveal the possible sites of functional change. A change in m means that the number of quanta of transmitter released is altered; that is, the presynaptic release mechanism is the site of change. A variation of $\bar{q}$ can be caused either by a presynaptic process (such as a change in the number

---

B: Isolated reflex test system. *B1*: Diagram of the experimental arrangement. The gill, siphon, and abdominal ganglion have been removed from the animal. The siphon skin is pinned to a lucite stage and stimulated with a mechanically driven constant-force probe having a tip diameter of 0.5 mm. The gill response is recorded with a photocell; intracellular recordings are obtained from L7. *B2*: Simultaneous records of gill contraction, the complex EPSP in L7, and the constant-force probe. L7 was hyperpolarized by about 50 mV.

C: Isolated ganglion test system. *C1*: Diagram of the isolated ganglion. The siphon nerve, which innervates the siphon, is electrically stimulated. Stimulation of the right connective (carrying fibers from the head) provides input for heterosynaptic facilitation and sensitization, to be considered below. Intracellular recordings from L7. *C2:* An EPSP is produced in L7 by a 1.5-msec electrical shock to the siphon nerve. The cell was hyperpolarized about 50 mV to prevent firing. The general configurations and time courses of the PSPs produced with natural stimulation of the gill (*A2, B2*) and with electrical stimulation of a nerve (*C2*) are roughly comparable. (Modified from Carew et al., 1971; Byrne et al., 1978.)

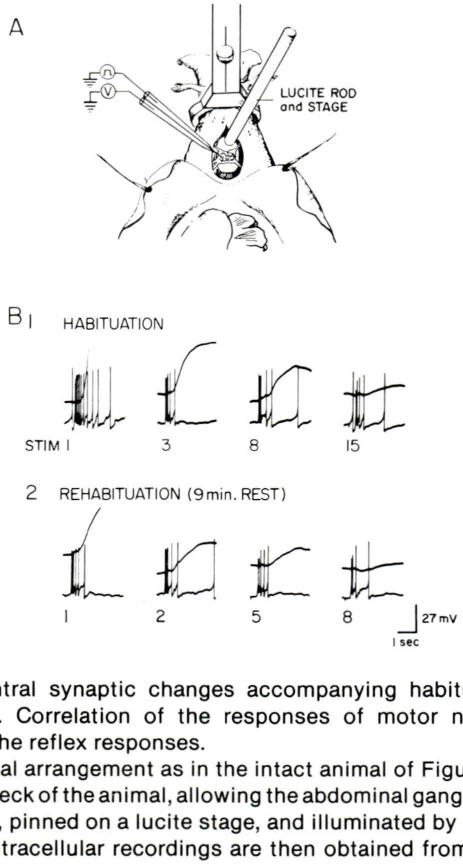

FIGURE 11. Central synaptic changes accompanying habituation and spontaneous recovery. Correlation of the responses of motor neuron L7 and of the amplitude of the reflex responses.

A: Experimental arrangement as in the intact animal of Figure 10*A*. A slit has been made in the neck of the animal, allowing the abdominal ganglion and its nerves to be externalized, pinned on a lucite stage, and illuminated by means of a lucite-rod light guide. Intracellular recordings are then obtained from identified motor cell L7 during gill withdrawal, allowing one to compare behavioral responses with motor neuron activity.

B: Sample records of gill contractions, measured with a photocell, and simultaneous intracellular recordings from identified motor neuron L7. Tactile stimuli (500 msec in duration) were presented to the mantle shelf every 90 sec. *B1*: Fifteen stimuli were presented over a period of 21 minutes. The number of action potentials in the first second following the first evoked action potential in each trace is, left to right, 9, 6, 6, and 4. *B2*: Rehabituation of the reflex following partial recovery due to a 9-min rest. The number of action potentials in the first second is 7, 6, 5, and 3. (From Kupfermann, Castellucci, Pinsker, and Kandel, 1970.)

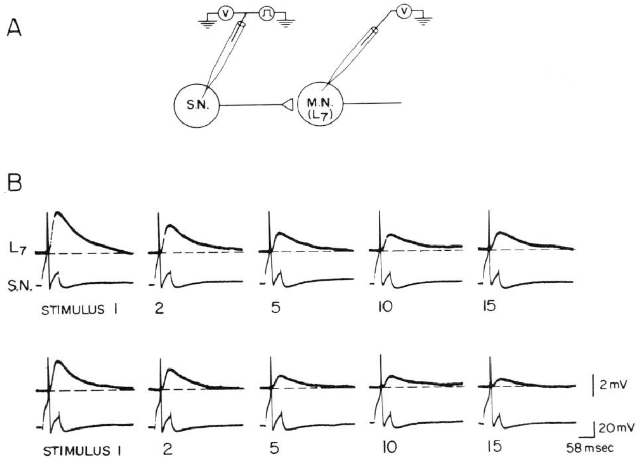

FIGURE 12. Depression of the monosynaptic EPSP produced in L7 by stimulating an individual sensory neuron.

A: Schematic drawing of the experimental arrangement showing the direct connections between the sensory neuron and motor neuron L7.

B: Changes in the effectiveness of the monosynaptic connection between the sensory and motor neurons as a result of repeated stimulation every 10 sec. Selected records from two consecutive training sessions consisting each of 15 consecutive stimuli. The second training session following a 15-min rest produces a more profound depression. The ganglion was bathed in solutions with high Mg^{++} and high Ca^{++} concentrations to block interneuronal activity. (From Castellucci and Kandel, 1974.)

of transmitter molecules packaged into each synaptic vesicle) or by a postsynaptic process (a change in receptor sensitivity or receptor density). On the other hand, a stable quantal size ($\bar{q}$) indicates that the sensitivity and density of the postsynaptic receptors are unaltered.

Figure 13 summarizes the results of a quantal analysis carried out by Castellucci and myself (1974) on synaptic depression at the sensory-to-motor neuron synapse. We found that as the average amplitude of the EPSP decreased to 50% of control with the repeated training that accompanies habituation, it was paralleled by a decrease in quantal output (m). By contrast, the quantal size did not change significantly.

The results of this quantal analysis are summarized in the diagram of Figure 14*A*, which illustrates that the *locus* of short-term habituation is presynaptic, involving a specific set of synapses: the synapses between

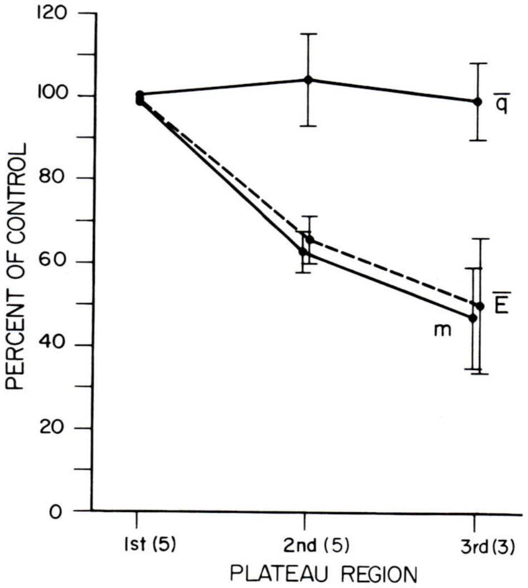

FIGURE 13. Comparison of an average EPSP amplitude ($\bar{E}$), quantal size ($\bar{q}$), and quantal output (m) during repeated stimulation (interstimulus interval, 10 sec). $\bar{E}$ is measured directly; m can be obtained in a number of ways. One way, illustrated in this experiment, is by means of the method of failures, whereby only the total number of stimuli and the number of failures in response need be known. This way of estimating m is derived from the limiting value of the Poisson distribution ($P_o = e^{-m}$) where P_o is the probability of producing a failure on any single trial. This probability is equal to the ratio of the number of failures following presynaptic stimulation (N_o) to the total number of stimuli (N). Thus, $m = 1n\,(N/N_o)$. Since $\bar{E} = m\bar{q}$, knowing $\bar{E}$ and m, one can now estimate $\bar{q}$.

Each point on the graph represents the mean of three to five experiments, as indicated in the parentheses. In each experiment, three consecutive plateau regions were obtained. These plateau regions are consecutive stable regions of about 50-100 EPSPs each, in which the EPSP depression is less than 15%.

During the consecutive plateau regions, the mean quantal output per stimulus (m) decreases in parallel with the decrease of the mean EPSP amplitude ($\bar{E}$). By contrast, the mean quantal size ($\bar{q}$) does not change significantly. (Modified from Castellucci and Kandel, 1974.)

the sensory neurons and their central target cells. The *mechanism* of habituation is a plastic change in the terminals of the sensory neurons: a transient homosynaptic depression. This depression results from a progressive decrease in the amount of transmitter released. Similar studies in the crayfish by Zucker (1972*b*) and by Krasne (1969, 1978) indicate that this mechanism and perhaps even this locus may be quite general (Figure 14*B*). I would caution, however, that given the variety

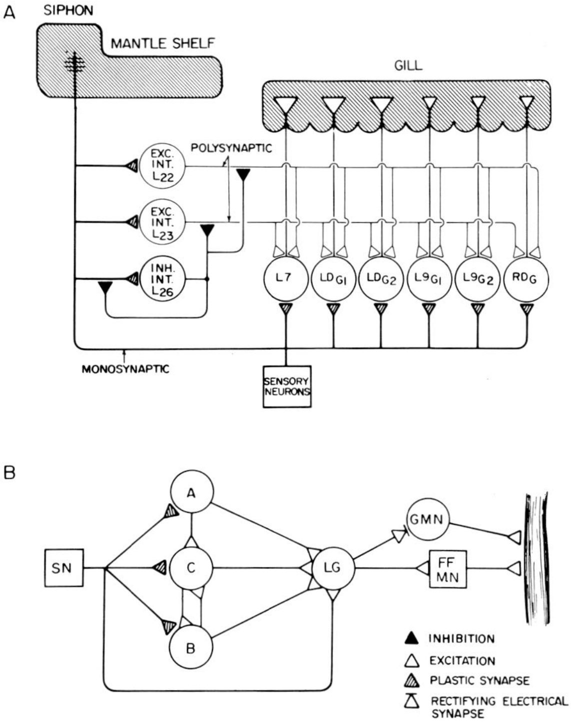

FIGURE 14. The locus and mechanism of habituation in *Aplysia* and crayfish.

A: Sites of neuron plasticity that mediate habituation in *Aplysia*. Schematic illustration of the identified neurons involved in the defensive gill-withdrawal reflex elicited by weak tactile stimulation of the siphon. All motor neurons and interneurons receive excitatory synaptic input from the sensory neurons. These excitatory connections have plastic capabilities (hatched) and undergo depression with repeated stimulation. The box represents a cluster of about 24 sensory cells. (Based on Castellucci and Kandel, 1974.)

B: Neural circuit for generating the tail-flip response of the crayfish in response to phasic mechanical stimulation of the abdomen. Circles represent single neurons; squares represent populations of similar neurons. Shaded terminals are chemical synapses that undergo depression with repeated stimulation. All contacts onto the lateral giant (LG) are electrical junctions. The electrical synapse between the lateral giant and the giant motor neuron (GMN) is unidirectional (rectifying). SN, tactile sensory neuron; A, B, C, identified tactile interneurons; LG, lateral giant; GMN, giant motor neuron; FFMN, fast flexor motor neurons. (From Zucker, 1972*a*.)

of forms that habituation can take in different response systems and in different animals, it seems likely that more than one mechanism exists.

What we have learned about the mechanism of habituation is consistent with the idea that, although genetic and developmental processes determine the connections between neurons (Sperry, 1951, 1963), the strength of the connections is only partially specified. It is this factor—the efficacy of synaptic connections—that is played upon by environmental factors such as learning.

Analysis of Long-Term Habituation

What are the limits of this plasticity? How much can the effectiveness of a given synapse change as a result of natural stimulation and how long can such changes endure? I had previously mentioned that repeated training sessions can depress the EPSP completely. Can this be maintained? Can long-term habituation produce a complete and prolonged inactivation of a previously functioning synapse?

These questions bear upon the long-standing debate, among students of learning, about the relationship of short-term and long-term memory. The most common notion, based on rather indirect evidence, is that short- and long-term memory involve different memory processes at different loci: a short-term dynamic change involving the reverberation of impulses in closed neuronal loops, and a long-term structural change (see for example Hebb, 1949; Agranoff, 1967; Barondes, 1970). However, as Weiskrantz (1970), Wickelgren (1973), Meyer and Beattie (1977), and others have argued, the available facts could equally well be explained by two (or more) phases of a single process. Since our results clearly showed that short-term memory could itself be mediated by a plastic change—and did not require the reverberatory process thought to be necessary for two processes of memory—the gill-withdrawal reflex seemed a particularly good system in which to examine the applicability of the one-process hypothesis to an instance of long-term memory. We therefore asked how short-term habituation becomes converted to the long-term process. Do both forms of habituation involve aspects of a common process? Do they involve the same locus—the connections made by the sensory neuron on its central target cells? Or are the two types of memory sufficiently distinct that different loci are involved?

To carry out this analysis, Castellucci, Carew, and I (1978) compared the connections between the sensory neurons and motor cell L7 in four groups of animals: (1) controls (untrained animals), (2) animals examined 1 day after long-term habituation training, (3) animals examined 1 week after training, and (4) animals examined 3 weeks after training. We found that in the control animals about 90% of sensory neurons made detectable connections to the major motor neuron L7. In contrast, in long-term habituated animals tested 1 day and 1 week after training, the percentage of detectable connections was reduced to about 30%. The remaining connections had been inactivated to the degree that they could not be clearly resolved electrophysiologically (Figures 15, 16). Even when tested 3 weeks later, only about 60% of the cells made detectable connections to L7 (Figure 16). Here then are functioning synaptic connections that become and remain depressed for over 1 week as a result of a simple learning experience consisting of several brief sessions of habituation training of 10 trials each.

Thus, whereas short-term habituation involves a transient decrease in synaptic efficacy, long-term habituation produces a more prolonged and profound change, leading to a functional inactivation of many of the previously existing connections (Figure 17). These data are interesting for two reasons: First, the data provide direct evidence that a specific instance of long-term memory can be explained by a long-term change in synaptic effectiveness. Second, the data show that at synapses critically involved in learning, the amount of training—the total number of train-

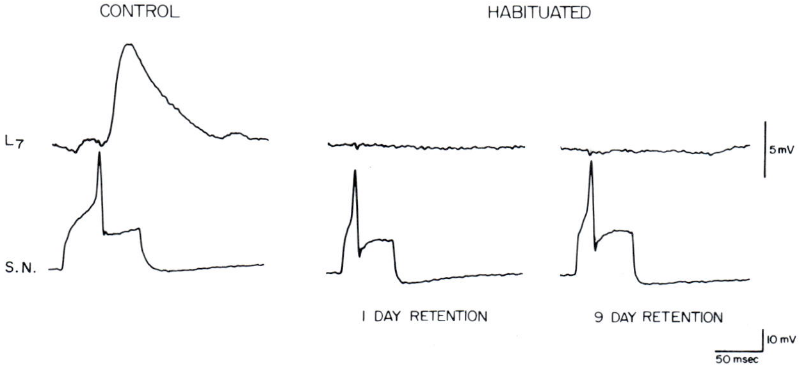

FIGURE 15. Comparison of a synaptic connection between a sensory neuron and the motor neuron L7 in control (untrained) animals and in animals that have been subjected to long-term habituation training and tested 1 day or 9 days after that training. The synaptic connection in trained animals is undetectable. (From Castellucci et al., 1978.)

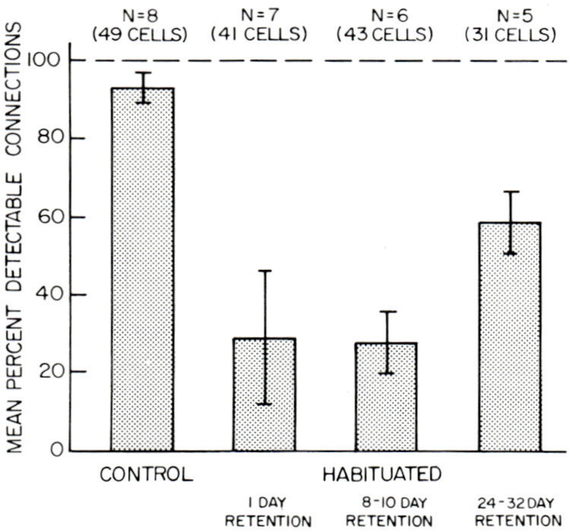

FIGURE 16. The mean percent of detectable connections in control and habituated animals tested at three intervals: 1 day, approximately 1 week, and approximately 1 month after long-term habituation training. The error bars indicate the standard error of the means. (From Castellucci et al., 1978.)

ing trials—necessary to produce a profound change in synaptic transmission is surprisingly small.

These studies also make clear that short- and long-term habituation share a common neuronal locus: the synapses that the sensory neurons make on the motor neurons. Short- and long-term habituation also involve aspects of the same cellular mechanism: depression of excitatory transmission. We now plan to do a quantal analysis to determine whether the long-term synaptic depression is pre- or postsynaptic. If long-term depression proves also to be presynaptic, it would support on a more fundamental level the behavioral notion that short- and long-term memory can involve a single memory trace.

How might such a single trace notion work? Might short- and long-term habituation involve a progressive retraction of the terminals or a change in the size of the terminals? Or might the change be more subtle morphologically? Among the more subtle changes, two possibilities come to mind (Figure 18). One possibility is that the short-term process results from the depletion (by inadequate mobilization) of a releasable store of transmitter, and the long-term process results from a depletion of the storage pool of transmitter by an inhibition of the synthesis of the transmitter-containing vesicles or their transport from the cell body.

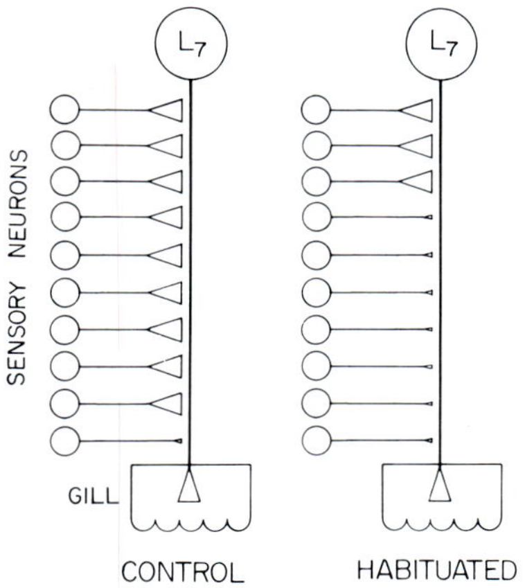

FIGURE 17. Hypothetical scheme illustrating the change in the functional effectiveness of synaptic connections (indicated by the size of the triangle symbolizing the synaptic terminal) between a population of 10 sensory neurons and motor neuron L7. In control animals, 9 out of 10 sensory cells make detectable connections on the motor cell L7 as indicated by large synaptic terminals. After undergoing long-term habituation training, only 3 of the 10 connections are detectable. The others have become functionally inactivated and their response is no longer detectable in the postsynaptic cell. This is indicated by small terminals. The difference in sizes of the triangles symbolizing the synaptic terminals is used only to illustrate functional changes in synaptic strength; the differences are *not* meant to imply a specific morphological mechanism.

Alternatively, there might be no change in the store of transmitter, the critical change occurring in the probability of release. The short-term process might be a transient decrease in the probability of release due to a transient decrease in Ca^{++} conductance in the terminals of the sensory neurons, whereas the long-term process could involve a prolonged shutting down of the Ca^{++} channels.

In the absence of data on long-term habituation and for reasons that I hope will become apparent later, I have recently developed a private bias for the second model. One direct approach for distinguishing between the two models, which Bailey, Thompson, Castellucci, and I (1978) have begun to use, is to label and visualize the terminals of the sensory neurons with the electron microscope after injection of horseradish peroxidase, whose reaction product is electron-dense and therefore a good intracellular marker (Muller and McMahan, 1976).

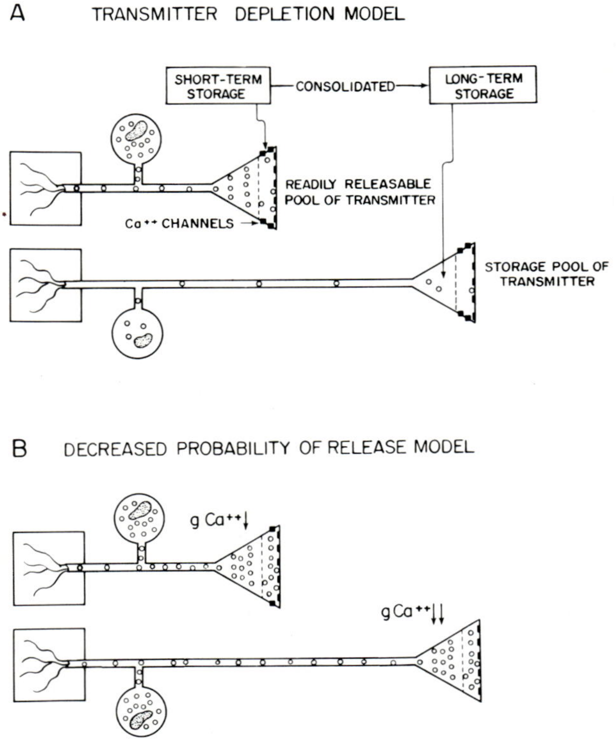

FIGURE 18. Two types of cellular models of a two-stage memory trace assuming a single cellular locus and a common plastic mechanism, synaptic depression.

A: Transmitter depletion model. In this model, short-term storage is assumed to involve mobilization of transmitter substance from a storage pool of vesicles to a readily releasable pool of vesicles. A single training session leads to partial depletion of the readily releasable pool (short-term habituation). Repeated training sessions produce depletion of the storage pool, which in turn could lead to reduced synthesis and transport of transmitter substance (long-term memory). According to this model, Ca^{++} conductance is not altered by either process, but vesicle number and distribution are altered.

B: Decreased probability of release model. In this model, both short- and long-term storage involve alterations in the probability of release resulting from a reduction of the Ca^{++} conductance in the terminals. A single training session leads to a transient reduction in Ca^{++} conductance (short-term memory), whereas repeated training sessions lead to a prolonged depression in Ca^{++} conductance (long-term memory). According to this model, vesicle number and distribution are only slightly altered by either process or not at all.

As a first step we examined the axonal tree of injected sensory neurons with the light microscope in 20-μm thick sections. The neuron was found to arborize and give rise to fine branches, along which are located an assortment of swellings or varicosities, 1 to 4 μm in diameter, interconnected by small diameter segments of variable length. Ultrastructural analysis of random slices taken through these varicosities reveals that they are filled with slightly elongate, smoothly contoured vesicles having a mean long axis of about 80 nm (Figure 19). When viewed in regions where horseradish peroxidase reaction product does not obscure their morphology, most of the vesicles appear to be relatively electron-lucent. The fine interconnecting segments between varicosities contain microtubules that are oriented parallel to the long axis of the segments (Figure 19). The vesicle-filled varicosities have regions of contact specialization with adjacent processes (Bailey, Thompson, Castellucci, and Kandel, 1976, 1978; Bailey and Thompson, 1978). On the basis of several criteria, including focal accumulation of synaptic vesicles, straightening and thickening of apposed membranes, and widening of the synaptic cleft, Bailey et al. (1978) have suggested that these contact specializations are active zones for synaptic transmission. Preliminary analyses of single sensory neurons suggest that the total number of varicosities per cell is roughly 450 and that each varicosity usually has 2 active zones. Thus, based upon these preliminary observations, a rough estimate for the total number of active zones per sensory neuron would be about 900. If we assume that each sensory neuron connects to 20 follower neurons, each follower would receive 45 active zone contacts from each sensory cell. Since a 5 mV synaptic potential characteristic of the unhabituated state has a quantal content of about 100, each active zone would, on the average, release 2 quanta per impulse.

Analysis of a limited number of completely reconstructed active zones between sensory neuron varicosities and postsynaptic processes has also begun to show a number of regularities on the fine structural level. The size of the active zone is positively correlated with the area of surface membrane belonging to and lying in a band around the postsynaptic element opposite the active zone: the larger the postsynaptic element, the larger the zone of specialization shared with the presynaptic varicosity. In addition, the number of vesicles touching the membrane per square micron of active zone decreases with increasing area of the specialization and with increasing size of the postsynaptic element (Bailey et al., 1978). These structural regularities make it reasonable to assume that

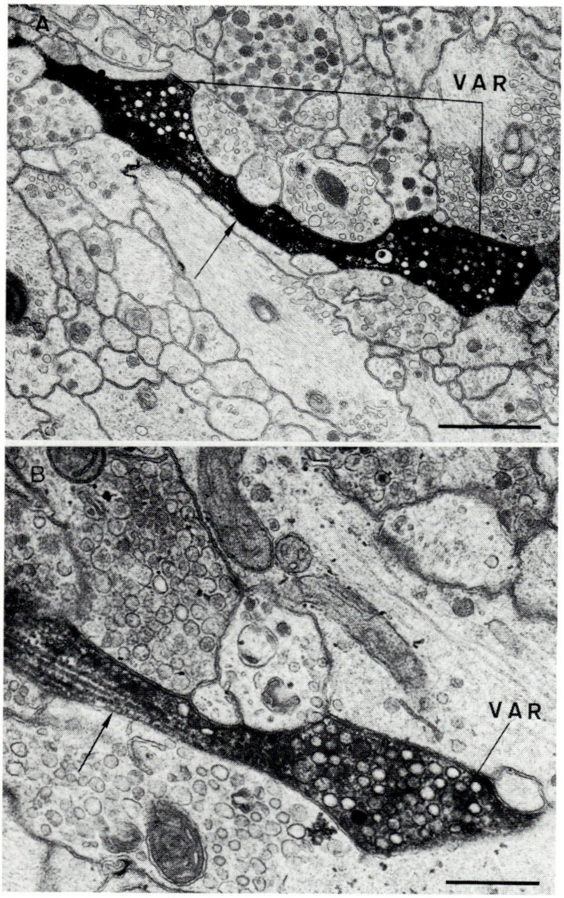

FIGURE 19. Fine structure of the sensory varicosity.

Top micrograph: This electron micrograph illustrates the appearance of two sensory varicosities and that of the intervaricosity segment as labeled with horseradish peroxidase. Each varicosity (VAR) is filled with vesicles, many of which appear to be electron-lucent, lying in a blanket of electron-dense reaction product. Scale = 1.0 μm.

Bottom micrograph: Detail of varicosity coming in apposition with typical small postsynaptic spine. The major intracellular component of the intervaricosity segment (arrows), well seen in this micrograph, consists of microtubules. Scale = 0.5 μm. (From Bailey et al., 1978.)

we might be able to detect significant changes in the size of the active zone or in the number or distribution of the synaptic vesicles (the likely storage sites of transmitter quanta) relative to these zones, were such changes to be associated with either short- or long-term habituation. Thus, the ability to examine in detail the morphological characteristics of these inactivated synapses provides the first direct opportunity to examine the structural basis of a long-term memory trace.

SENSITIZATION: A MORE COMPLEX FORM OF LEARNING

I have so far described a very elementary form of learning whereby reflex activity is depressed. To what degree can one now apply a cellular strategy to a slightly more complex learning paradigm? The next step in complexity is a form of learning called *sensitization.*

Sensitization is the enhancement of an animal's preexisting reflex response by a strong or noxious stimulus. This enhancement can last from minutes to weeks depending upon the pattern and duration of training (Figures 20, 21). Because both sensitization and arousal lead to increased responsiveness that is generalized and sustained, sensitization is thought to be a critical component of behavioral arousal (Groves and Thompson, 1970; Carew, Castellucci, and Kandel, 1971; Kandel, 1978).

Sensitization is more complex than habituation and resembles classical conditioning in that activity in one pathway facilitates reflex activity in another. Unlike classical conditioning, however, the reflex facilitation caused by sensitization does not require specific temporal association of the two stimuli. In addition to being of interest in its own right, sensitization is also a means for enhancing habituated responses. Earlier studies by Spencer and his colleagues on the flexion reflex in the cat spinal cord (Spencer et al., 1966*a,c*) and our more recent studies on *Aplysia* (Carew et al., 1971) have shown that what is called dishabituation—the enhancement of a habituated response by a strong stimulus—is only a special example of sensitization. The same stimulus will also enhance nonhabituated control responses. As is the case for habituation, sensitization has both a short- and a long-term form (Figure 21; Pinsker et al., 1973). Here, however, I will focus only on the short-term form.

Physiological Analysis

At the cellular level, sensitization involves an alteration of synaptic transmission at the same locus that is involved in habituation: the

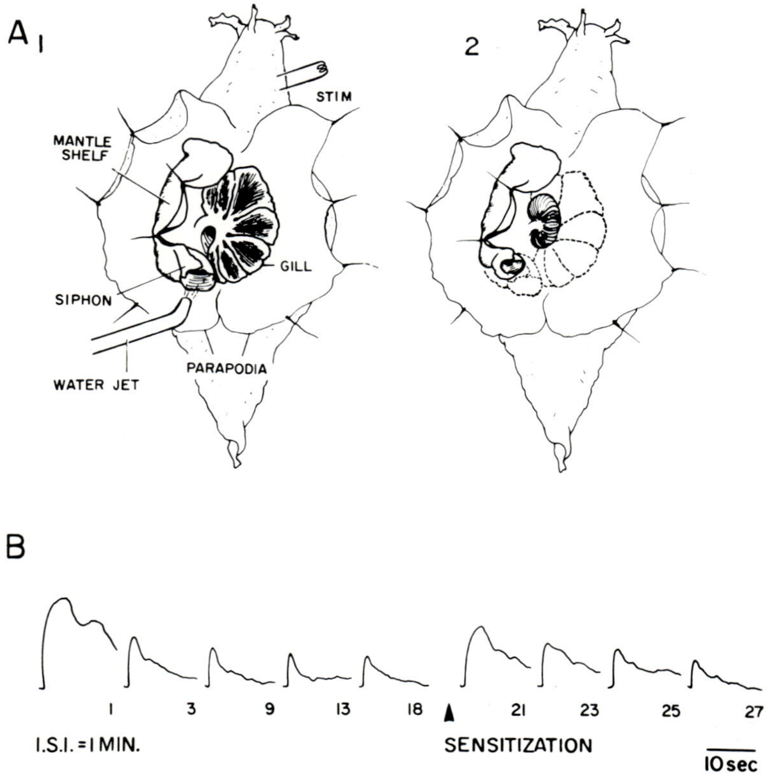

FIGURE 20. Short-term sensitization of the gill-withdrawal reflex in *Aplysia*.

A: Experimental arrangement for the behavioral studies in the intact animal. A gill-withdrawal reflex is elicited by a water jet (tactile) stimulus to the siphon. The sensitizing stimulus is a noxious mechanical or electrical stimulus to the head or neck.

B: Habituating stimuli were presented to the siphon at 1-min intervals. A noxious mechanical stimulus was applied to the neck. Following this sensitizing stimulus, the response was facilitated for several minutes. (Pinsker et al., 1970.)

synapses made by the sensory neurons on their central target cells. Sensitization can be produced in the intact animal by noxious stimulation applied to the head or another part of the body; it can also be produced in the isolated abdominal ganglion by stimulating the connectives, the pathways from the head that mediate sensitization (Figures 22, 23; Castellucci et al., 1970; Hawkins, 1978). Following a sensitizing stimulus, synaptic transmission is enhanced, and a quantal analysis reveals that this enhancement results from an increase in the amount of transmitter released by each action potential in the sensory neuron (Figure 24; Castellucci and Kandel, 1974).

These findings, summarized in Figure 25, have led us to suggest that sensitization involves a presynaptic facilitation in a neural pathway whereby terminals from the pathways mediating sensitization end on the terminals of the sensory neuron and modulate its transmitter release. Thus, our data indicate that the same locus can be regulated in two ways: it can be depressed (homosynaptically) by habituation and it can be facilitated (heterosynaptically) by sensitization.

Recently Hawkins, Castellucci, and I (1976, 1978*a*) have started to analyze the pathway from the head that produces presynaptic facilitation. We have identified two types of interneurons in the abdominal ganglion—cell L28 and cell group L29, a cluster of electrically interconnected cells. Cell L28 and the L29 group are higher order excitatory interneurons in the gill-withdrawal reflex. One of them receives a direct connection from the sensory neurons and connects indirectly to the motor neurons; the other receives an indirect connection from the sensory neurons and makes a direct connection to the motor neurons (Figure 27*A*). Both cell types are fired very effectively by stimulating the connective—the pathway from the head that mediates sensitization. Firing a facilitator produces little synaptic activity as recorded in the cell body of either the motor cell or the sensory cell, but greatly enhances the effectiveness of the synaptic actions of the sensory neurons (Figure 26).

As might be expected from a modulatory neural circuit involved in arousal, the neural circuit for presynaptic facilitation has extensive actions and affects a number of closely related pathways (Carew et al., 1971). Thus, as indicated in the schemes of Figure 27*A* and *B*, behavioral habituation and its cellular mechanism, homosynaptic depression, are restricted to the particular sensory neurons whose receptive field is stimulated. By contrast, behavioral sensitization and its cellular aspect, presynaptic facilitation, are broadly distributed and will facilitate responses elicited from widely different points on the receptive field, such as the siphon and the mantle shelf, points innervated by independent clusters of sensory neurons (Figure 7).

An interesting feature emerges from an examination of the simplified wiring diagrams of Figure 27. The diagrams illustrate an interaction between the two parallel and overlapping systems: (1) a mediating system that initiates the gill-withdrawal reflex, for simplicity illustrated here

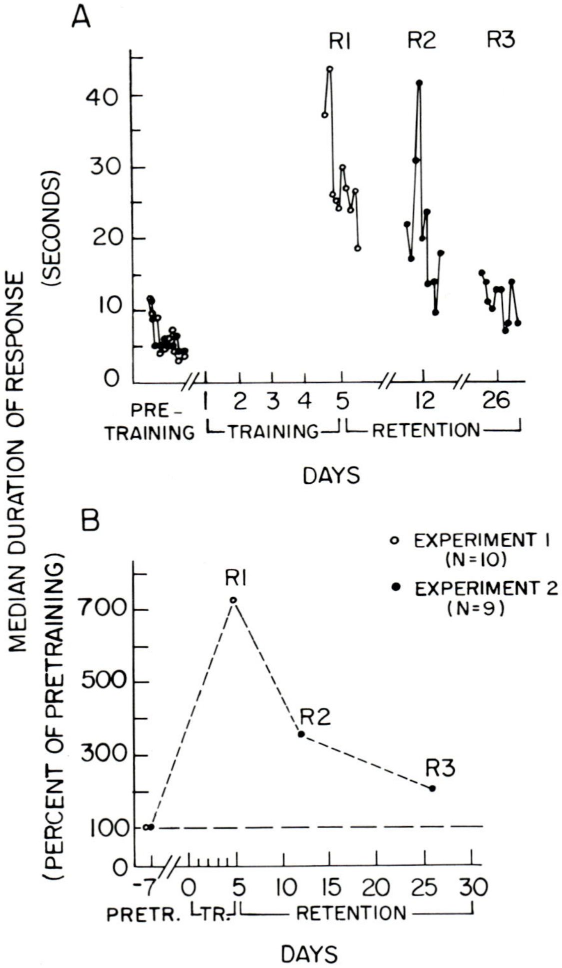

FIGURE 21. Long-term sensitization of siphon withdrawal.

A: The median duration of siphon withdrawal is shown for each trial of a 10-trial block (minimum intertrial interval was 30 sec). The results of two independent experiments are shown. Experimental animals were given four electrical shocks per day for 4 days; the controls received no shocks. In experiment 1, retention of sensitization was tested 1 day (R1) after the last shock. In experiment 2, retention was tested 1 week (R2) and 3 weeks (R3) after the last shock. Significant sensitization was evident in the 1-day and 1-week retention tests, while there was almost complete recovery in the 3-week retention test.

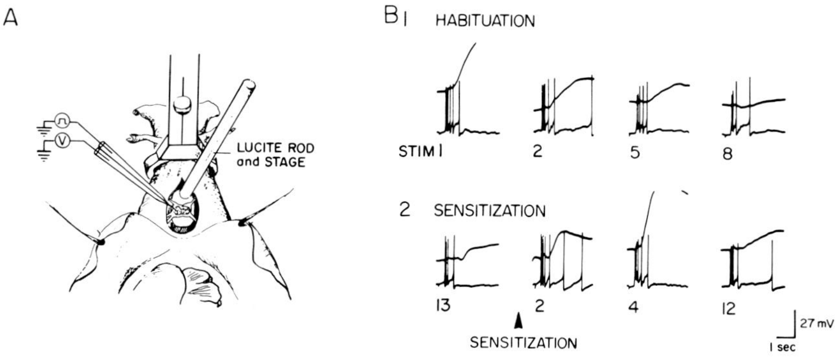

FIGURE 22. Central synaptic changes accompanying sensitization. Correlation of the responses of motor neuron L7 and of the amplitude of the reflex responses.

A: Experimental arrangement as in Figure 11. A slit has been made in the neck of the animal, allowing the abdominal ganglion and its nerves to be externalized. Intracellular recordings are then obtained from identified motor cell L7 during gill withdrawal, allowing one to compare behavioral responses with motor neuron activity.

B: Sample records of gill contractions, measured with a photocell, and simultaneous intracellular recordings from identified motor neuron L7. Tactile stimuli (500 msec in duration) were presented to the mantle shelf every 90 sec. *B1*: This is the same experiment illustrated in Figure 11*B*, showing the second habituation session. The number of action potentials in the first second following the first evoked action potential in each trace is 7, 6, 5, and 3. *B2*: Following the last habituation trial shown in the first trace, a strong stimulus was applied to the animal. The discharge of the motor neuron and the amplitude of the gill contraction progressively increased during the first three stimuli following the sensitizing stimulus and remained elevated for several minutes. The number of action potentials in the first second following the first action potential in each trace is 4, 5, 7, and 5. (From Kupfermann et al., 1970.)

B: Time course of sensitization (data from part *A*). Responses for each daily session have been summed and expressed as a single (median) score. Twenty-four hours and one week after the end of training, the experimental animals showed significantly longer duration siphon withdrawal compared to their own pretraining score and the score of the controls, both of which are expressed in the graph as 100% (Mann-Whitney *U* tests used for intergroup comparisons). In the 3-week retention test there was no longer a significant difference between the experimental and control groups, but the responses of the experimental animals were still significantly prolonged compared to their own pretraining levels (Wilcoxon matched pairs signed-ranks test: $p < 0.005$); the control animals were unchanged. (From Pinsker et al., 1973.)

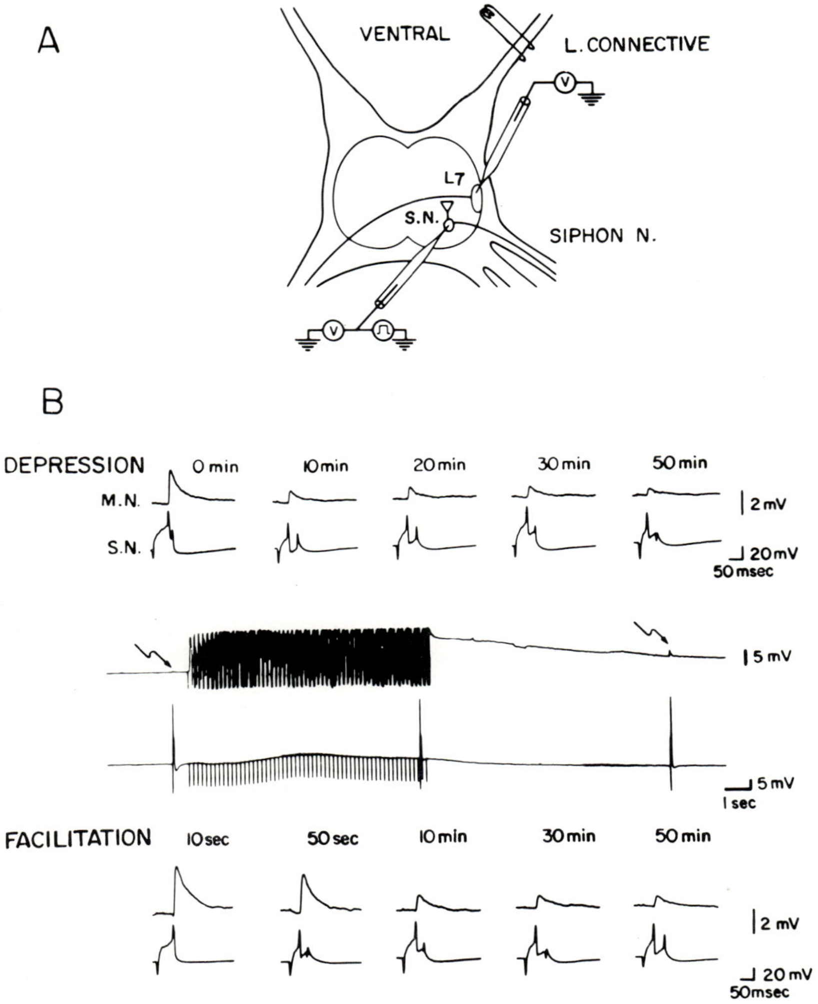

FIGURE 23. Synaptic facilitation at the synapse between mechanoreceptor neurons and motor neurons.

A: Ventral aspect of the abdominal ganglion of *Aplysia*, illustrating simultaneous recording from gill motor neuron L7 and a mechanoreceptor sensory neuron (SN). Stimulation of the left connective, which carries information from the head to the abdominal ganglion, is used as the facilitating stimulus.

B: Depression and subsequent facilitation of the monosynaptic EPSP after a strong stimulus. Top two series of traces illustrate selected records of depres-

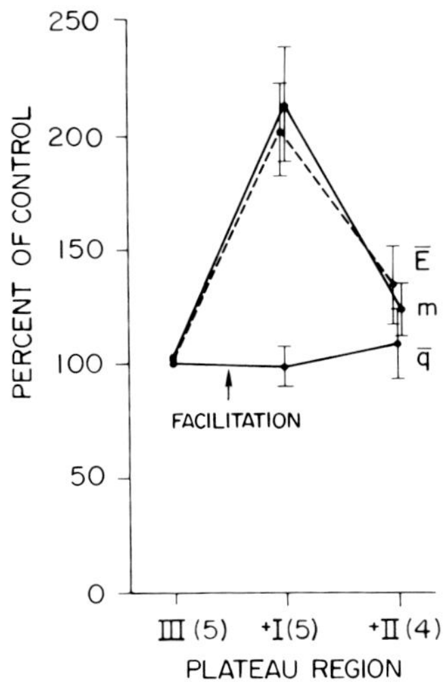

FIGURE 24. Quantal analysis of synaptic facilitation. Data based on estimates of *m* and $\bar{q}$ derived according to the failure method (see also Figure 13 legend). The graph illustrates the plateau region preceding the facilitating stimulation (III) and two subsequent regions (+I and +II). The estimated values of $\bar{q}$ obtained in this way did not vary significantly between successive regions. By contrast, *m* paralleled $\bar{E}$. Both increased from a control value of 100% to a maximum value of 200% before returning with continued stimulation toward the control value. The numbers of experiments contributing to each point are indicated in the parentheses. (From Castellucci and Kandel, 1976.)

---

sion of the EPSP in the motor cell (MN) during a series of 50 min of consecutive intracellular stimuli to the sensory neuron (SN) (every 10 sec) prior to the presentation of the facilitating stimulus. Middle two traces illustrate the effects of the facilitating stimulus (a train of shocks to the LC at 6 Hz for 10 sec). Arrows indicate the last EPSP before the facilitating stimulus and the first EPSP after the stimulus. Bottom series of two traces illustrate the gradual decline of the facilitation over the subsequent 50 min, during which the sensory neuron was again stimulated once every 10 sec. The middle two traces are at lower gain and at a slower time scale. Abbreviations: SN, sensory neurons; MN motor neuron. (From Castellucci and Kandel, 1976.)

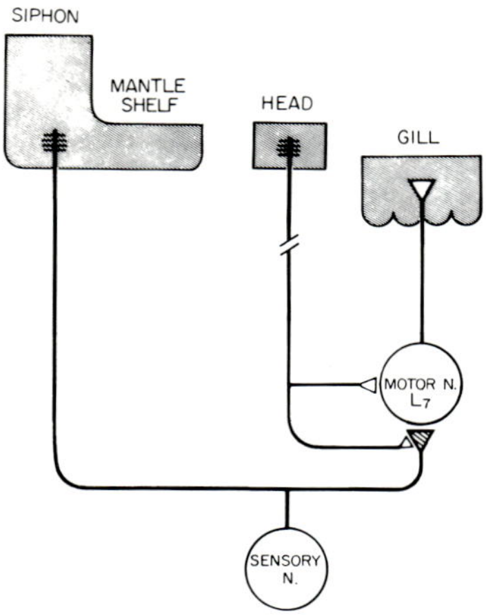

FIGURE 25. Postulated plastic changes in the circuit underlying habituation and sensitization of the gill-withdrawal reflex. For schematic purposes, only one sensory neuron (SN) and one motor neuron (L7) are shown. Similar processes are thought to be operative at all the synaptic contacts made by the population of sensory neurons on the other motor neurons and interneurons. The terminal of a sensory neuron (SN) is the common locus of both the synaptic depression underlying habituation and the presynaptic facilitation underlying sensitization. The pathway from the head mediating presynaptic facilitation is postulated to end on or near the sensory neuron terminal. The plastic synapse from the sensory-to-motor neuron, indicated by shading, can thus be regulated in two ways: it can be depressed by habituation and facilitated by sensitization. (From Kandel et al., 1976*a*.)

as a monosynaptic reflex, and (2) a modulatory system that controls the gain of the mediating system, illustrated here as the neurons producing facilitation. The function of the mediating circuit is to produce the behavior—a function it can exercise without the participation of the modulatory system. The main function of the modulatory system is not to initiate the behavior but to regulate its strength, which it does by enhancing transmission at certain critical synapses in the mediating system.

The study of sensitization thus reveals, on the cellular level, a specific instance of a commonly encountered behavioral case, the regulation of a mediating system by a modulatory one. The functioning of this modulating system is particularly interesting from a cellular point of view. Whereas we know a great deal about the mechanisms of synaptic trans-

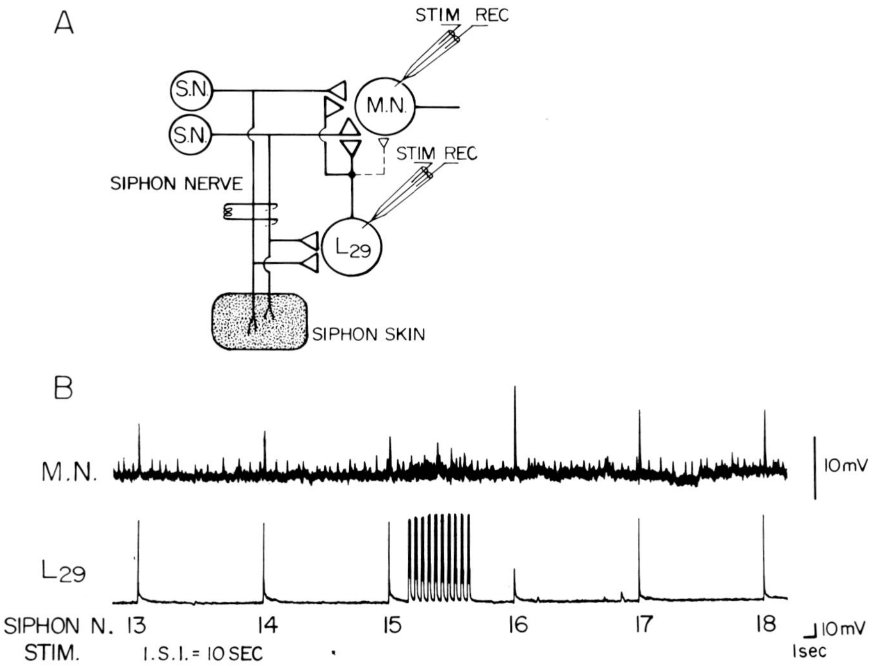

FIGURE 26. EPSP facilitation produced by intracellular stimulation of a single neuron.

A: The experiment arrangement. Intracellular recordings are obtained from a single motor cell. A population of sensory neurons (of which only two are illustrated schematically) is activated by stimulating their axons in the siphon nerve. A facilitatory neuron, one of the L29 group, is recorded from and stimulated intracellularly. L29 receives direct connections from the sensory cells and makes a weak indirect connection to the motor cell. In addition, L29 is postulated to make a direct connection to (or near) the terminal of the sensory neurons, leading to synaptic facilitation in that terminal.

B: EPSP depression by repeated siphon nerve stimulation and its facilitation by L29. The last three PSPs of a training session of 15 consecutive stimuli are shown before the facilitating stimulus. Top trace: The complex PSP produced in a gill motor neuron by stimulation of the axons of the sensory neurons in the siphon nerve at 10-sec intervals. Bottom trace: Intracellular recording from L29. This cell group also receives an EPSP (which produces a spike in all but trial 16) from the sensory neurons. Intracellular stimulation of L29 between trials 15 and 16 produces repetitive firing of interneuron L29 and a facilitated EPSP in the motor neuron. (From Hawkins et al., 1978*a*.)

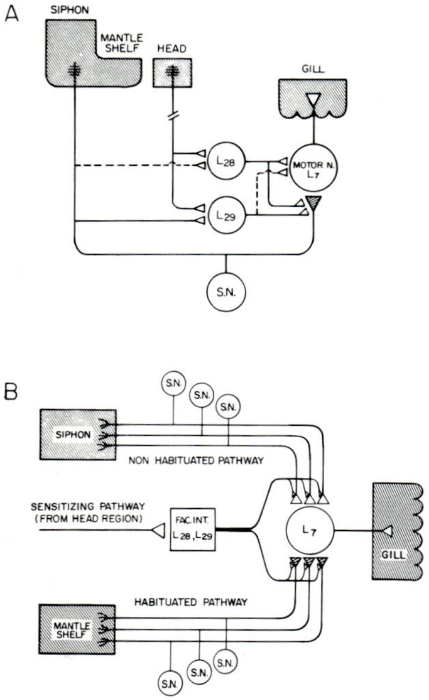

FIGURE 27. Interaction of the modulatory neurons producing facilitation with the neurons that mediate the gill-withdrawal reflex.

A: Modulation of synaptic transmission by individual identified cells. The pathway from the head that mediates sensitization ends on a group of interneurons, two of which—L28 and the L29 cells—have been identified. These neurons are postulated to end on the terminals of the sensory neuron, where they produce presynaptic facilitation of transmission between the sensory neuron and L7.

B: Schematic comparison of sensitization and habituation of the gill-withdrawal reflex. Habituation is selective and quite restricted in its action. Following stimulation of either the mantle shelf or the siphon, only the stimulated pathway is habituated; the homosynaptic (self-generated) depression is limited to synapses (hatched triangles) between the activated sensory neurons (SN) and motor neuron L7. By contrast, a strong stimulus from the head sensitizes both the habituated and nonhabituated pathways. This results from the fact that presynaptic facilitation acts on all the synapses made by primary sensory neurons on the motor neuron L7. The difference in the spatial distribution of the pathways mediating habituation and sensitization explains the restricted nature of habituation and the widespread action of sensitization. (From Carew et al., 1971.)

mission, we know little about gain control in synaptic pathways: how chemical transmission in one pathway is modulated for long periods of time by a chemical mediator (a transmitter or a hormone) released from another pathway. Yet it seems probable that much neural transformation in the brain involves the action of modulatory systems.

Because the action of modulatory systems of the sort involved in sensitization may prove a common mode of neural regulation, it will be of interest to examine further two of its aspects: (1) What are the fine structural features of the synapses of the modulatory neurons? Do the modulatory synapses form well-delineated active sites on the presynaptic terminals of the sensory neurons with closely apposed membranes and membrane specializations, or does the morphology suggest more diffuse transmitter release without specialized release sites, as appears to be the case in certain autonomic synapses? And, (2) how does the presynaptic neuronal action enhance transmitter release in the terminals of the sensory neuron? How do the mechanisms for synaptic modulation differ from those of conventional synaptic actions? I will first consider preliminary morphological studies before reviewing current studies designed to explore the molecular mechanisms underlying sensitization.

Steps Toward a Morphological Analysis

To obtain some information about the synapses that modulate the release of the transmitter substances from the terminals of the sensory neurons, Bailey, Thompson, Castellucci, and I have explored the structures that come into apposition with these terminals.

Unlike most other neurons in this ganglion, sensory neurons that mediate the gill-withdrawal reflex receive very little synaptic input. Consistent with this observation, Bailey et al. (1976, 1978) have found so far only two prominent types of processes contacting the dendritic tree of the sensory neurons injected with horseradish peroxidase. Both types of terminals contain dense-core granules, suggesting that the transmitter for the modulatory synapses may be a biogenic amine (Geffen and Livett, 1971; Lagercrantz, 1976). The first class contains vesicles with a mean long axis of 108 nm, many of which have a small, dense inner core (Figure 28). This class of presynaptic input forms active zones with sensory processes. The second class contains larger, diffusely dense-cored granules with a mean long axis of 123 nm (Figure 29). Unlike the first morphological class of synaptic profiles contacting the sensory neuron, however, the appositions of dense-core granule-containing

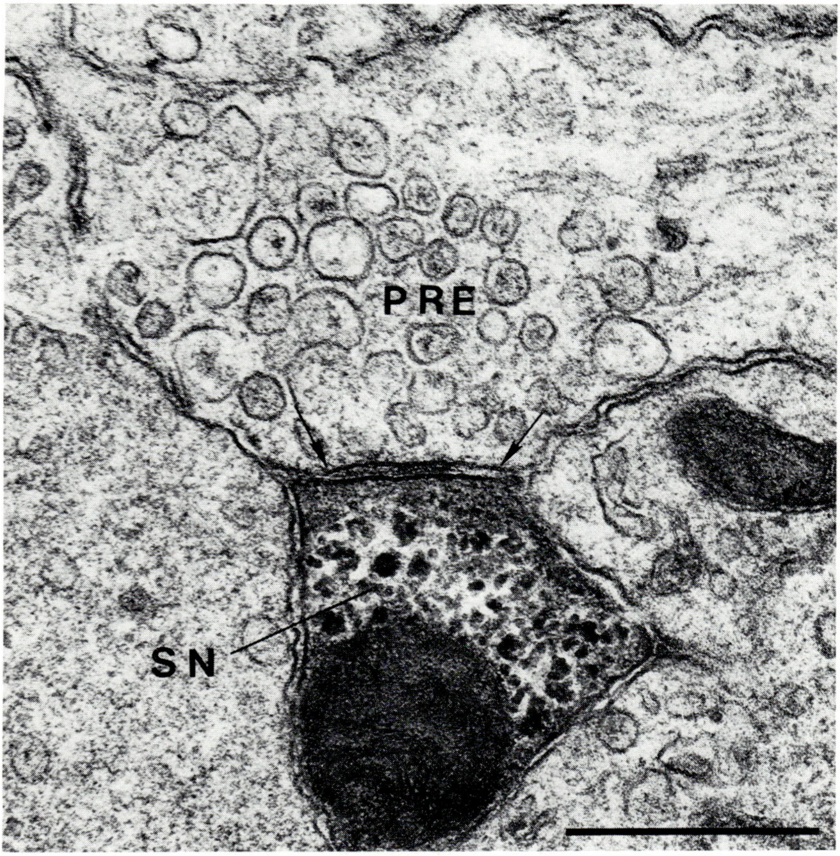

FIGURE 28. This electron micrograph illustrates an example of one of the two morphological classes of presumptive presynaptic input onto the sensory tree. This class forms active zones with sensory processes and contains vesicles many of which feature a small, dense inner core. A profile of this class of presynaptic (PRE) terminals is shown here making synaptic contact (between arrows) onto a small, HRP-filled sensory neuron process (SN). Scale = 0.5 μm. (From Bailey et al., 1978.)

processes lack any evidence of well-delineated active zones, such as widening of the synaptic cleft, membrane straightening or thickening. Even though it lacks an active site, this type has attracted our attention because of its ubiquitous occurrence. In the course of examining large

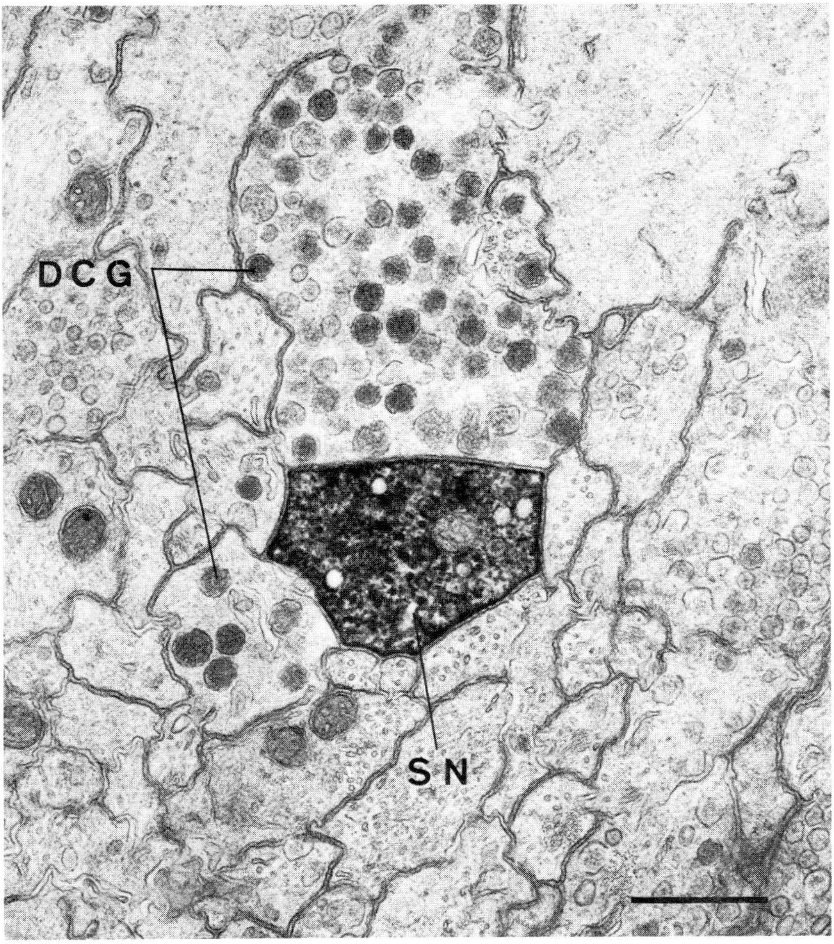

FIGURE 29. This micrograph illustrates a second and common structure that contacts the axonal tree of the sensory neuron in regions where varicosities occur. This type of profile is characterized by large, dense-core granules (DCG) and is frequently found touching sensory varicosities and intervaricosity segments. Such regions of contact do not have morphologically recognizable active zones. As is illustrated here, the apposition between dense-core granule-containing profiles and the sensory neuron varicosity (SN) can be extensive, with a large portion of the varicosity's perimeter in contact with dense-core profiles. Scale = 0.5 μm. (From Bailey et al., 1978.)

numbers of serial sections through sensory neuron varicosities filled with horseradish peroxidase, we almost invariably found this second type of process either touching the varicosity or the intervaricose segment, or nearby in the field.

As the first step in determining whether either one of these dense-core vesicle-containing profiles corresponds to the physiologically identified classes of facilitatory neurons, we have injected the L29 cells with horseradish peroxidase and compared their vesicle types with those found in the two morphological classes of processes that contact the sensory neuron. Our preliminary results suggest that cell L29 corresponds to the first class of processes, those containing vesicles with small, dense inner cores. Some of the labeled processes of L29 have clearly identified active zones apposed to unidentified processes in the neuropil. We now plan to examine the nature of the contacts between the facilitator neuron and the sensory neurons by labeling them separately with different substances, a method developed recently by Thompson, Bailey, Castellucci, and Kandel (1978).

Steps Toward a Molecular Analysis

A nice feature of short-term sensitization is that its chemical mechanisms may also be amenable to analysis. In collaboration with James Schwartz, Castellucci and I (1971) first examined the effects of protein synthesis inhibitors (such as anisomycin) and found that they did not interfere with sensitization, suggesting that short-term sensitization does not require the synthesis of new proteins. One possibility that now occurred to us was that sensitization may depend on a small molecule, perhaps an intracellular messenger like cAMP, that could regulate availability of transmitter substance within the presynaptic terminals. With this in mind, Cedar, Schwartz, and I (1972) examined cyclic AMP levels after stimulation of the connectives and found that strong and prolonged stimulation produced a synaptically-mediated increase in cyclic AMP in the whole ganglion.

Subsequent studies by Cedar and Schwartz (1972) and by Levitan and Barondes (1974) showed that this type of increase could be produced by incubating the ganglion with one of three biogenic amines: dopamine, octopamine, or serotonin. In each case these putative transmitters produced a prolonged increase in cyclic AMP. The time course of the elevation of cyclic AMP found by Cedar and Schwartz in the abdominal ganglion following 5-minute incubation with 10^{-4} M serotonin is compared with the time course of presynaptic facilitation to a very strong stimulus in Figure 30. The interesting aspect of this increase is that it is maintained for almost an hour and thus parallels the time course of presynaptic facilitation. This parallel in time course in turn suggested to us the idea that presynaptic facilitation is due to the action of one or more of the three biogenic amines, octopamine, dopamine, or serotonin (a possibility consistent with the morphology of the candidate modulatory synapses), and that the amines affect transmitter release by increasing the level of cyclic AMP in the presynaptic terminals of the sensory neuron.

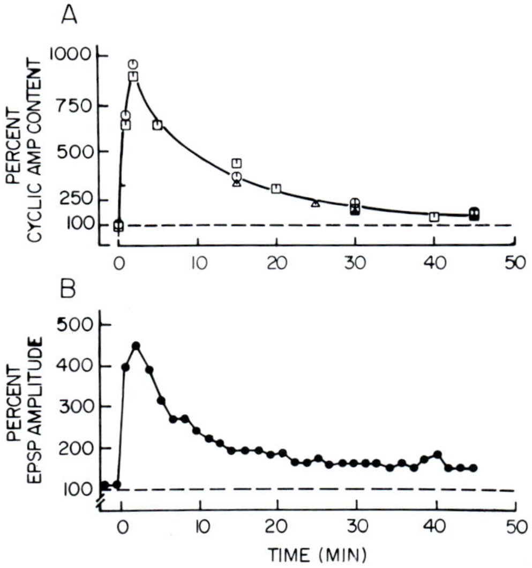

FIGURE 30. Comparison of the time course of the increase of cyclic AMP level in the abdominal ganglion and synaptic facilitation.

A: Increase of cAMP content. The abdominal ganglia from animals weighing an average of 130 g were incubated in 0.2 mM serotonin for 5 min. Cyclic AMP levels were determined at about 5-min intervals for almost an hour. Each value is from two or more determinations that were normalized to the amount of cAMP in the tissue 5 min after onset of incubation with serotonin. (Modified from Cedar and Schwartz, 1972.)

B: Presynaptic facilitation of a monosynaptic EPSP. A sensory neuron was stimulated once every 10 sec. The EPSP amplitudes have been normalized in relation to the average of the 20 EPSPs immediately preceding the facilitation (which was taken as 100%). Each point of the graph is the average of 10 successive EPSPs. The facilitating stimulus was electrical shocks to the left connective (6 Hz for 10 sec). (From Kandel et al., 1976*a*.)

We therefore examined the actions of the three amines on the monosynaptic connection between the sensory neurons and the motor cell L7 and found that one of them—serotonin—in a concentration as low as 1×10^{-6} M did indeed enhance synaptic transmission following a 2-minute exposure (Figure 31*A*). In addition, the enhancement produced by stimulating the connective or by stimulating a single facilitatory neuron—L28 or L29—is blocked by the serotonin-blocking agent cinanserin (Brunelli, Castellucci, and Kandel, 1976; Hawkins et al., 1976, 1978*a*).

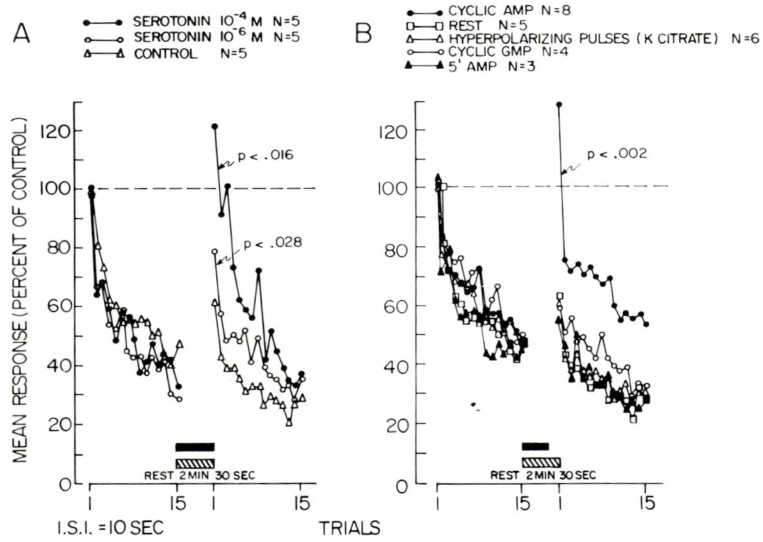

FIGURE 31. Facilitation of a monosynaptic EPSP by serotonin and cAMP. *A*: Effect of serotonin on monsynaptic EPSP. A sensory neuron was stimulated and produced a monosynaptic EPSP in a gill or siphon motor neuron; the interstimulus interval (ISI) was 10 sec. In each experiment, the EPSP amplitudes were normalized with respect to an initial control EPSP (N = 5, in all cases.) In the control group (open triangles), 15 stimuli produced synaptic depression. After a rest of 2½ min (hatched bar) and a slight recovery of EPSP, a second group of 15 stimuli was applied to the sensory neuron, producing further depression. In the two experimental groups, the 15 initial stimuli produced similar synaptic depression. During the 2½-min rest period (black bar), serotonin (2×10^{-4} M [closed circles] or 2×10^{-6} M [open circles]) was introduced into the perfusing solution. During the initial EPSP depression, the control and experimental curves overlapped completely. But after treatment with serotonin,

To determine whether serotonin might act by means of cyclic AMP, Brunelli, Castellucci, and I incubated the ganglion with dibutyryl cyclic AMP, an analog of cyclic AMP that crosses the cell membrane readily, and found that it enhanced synaptic transmission at the sensory neuron-to-motor neuron synapse but not at two other chemical excitatory synapses that we examined. But to obtain a more direct test, we injected cyclic AMP itself intracellularly into the cell body of the sensory neuron with hyperpolarizing current pulses through electrodes filled with 1.5 M cyclic AMP and tested the monosynaptic connections 2 minutes later. Intracellular cyclic AMP also produced facilitation, whereas cyclic GMP and 5′AMP (the breakdown product of cAMP) did not cause facilitation (Figure 31*B* and Brunelli et al., 1976). This enhancement can also be simulated by guanosine triphosphate (GTP), which has been shown in some systems (Rodbell, 1975) to stimulate directly adenylate cyclase, the enzyme that synthesizes cAMP (Klein, Castellucci, and Kandel, unpublished observation).

These data suggested a preliminary model of how presynaptic facilitation might work (Figure 32). The facilitation might be mediated by the terminals of serotonergic neurons, perhaps the processes containing the small, electron-dense granules that we considered above. We posit that these serotonergic terminals end on or near the terminals of the

---

both experimental curves were significantly higher than the control curve (triangles; $p < .016$ for 2×10^{-4} M and $p < .028$ for 2×10^{-6} M; one-tailed Mann-Whitney *U* test). To determine the statistical difference between the experimental and the control curves, the amplitudes of the responses to each of the 15 stimuli of a run were summed to obtain a single score.

B: The effect of cyclic AMP injection. The experimental group (closed circles, N = 8) is compared with a first control group, in which the rest (2½ min, indicated with hatched bar) was simply followed by 15 additional stimuli without any current injection (open squares, N = 5). In a second control group, a hyperpolarizing current pulse, comparable to that used to inject cyclic AMP, was injected into the sensory neuron, but the pulses were applied through an electrode filled with 3 M potassium citrate (open triangles, N = 6). In a third group, cyclic GMP (1.5 M) was injected intracellularly (open circles, N = 4). In a fourth group, 5′AMP (closed triangles, N = 3) was injected. Black bar indicates 2-min period of injection used in all experiments. The five curves generated during the initial EPSP depression produced by the first 15 stimuli overlap completely. But after injection of cyclic AMP (filled circles) the experimental group was significantly higher than either the first control group ($p < .002$, one-tailed Mann-Whitney *U* test) or the three other control groups (for potassium citrate, $p < .001$; for cyclic GMP, $p < .014$; for 5′AMP, $p < .012$). (From Brunelli et al., 1976.)

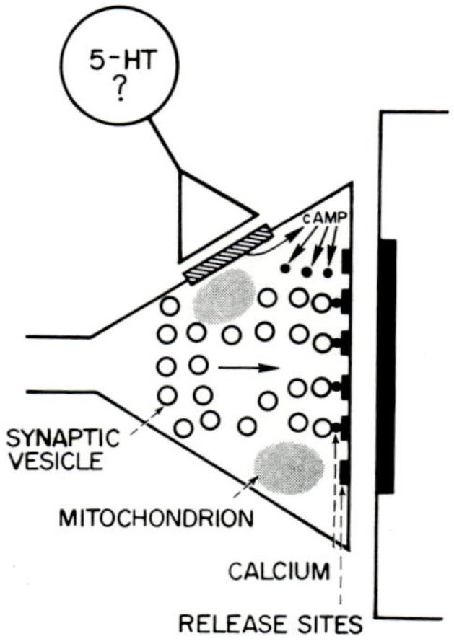

FIGURE 32. A preliminary model of sensitization. The presynaptic terminal of a sensory neuron is schematically illustrated with its complement of synaptic vesicles (small circles), mitochondria (shaded ovals), and active sites (thin black bars). The subsynaptic region of the postsynaptic neuron is also shown (thick black bar). The terminal of the postulated presynaptic neuron that releases serotonin at its terminal is indicated. The serotonin acts on a membrane-bound adenylate cyclase to increase cAMP in the terminals of the sensory neurons, which leads to an increase in transmitter release, perhaps by means of an increase in free Ca^{++}, indicated as black dots. (Modified from Kandel et al., 1976*a*.)

sensory neurons and act there to increase the concentration of cAMP in the terminals. Data in support of this aspect of the model have recently also been obtained for presynaptic facilitation at another synapse (Shimahara and Tauc, 1975).

The critical ion for transmitter release is Ca^{++}, which is thought to be responsible for allowing transmitter vesicles to bind to discharge sites—a precursor step in exocytotic release. Ca^{++} influx into the terminal is triggered by the depolarizing action of the action potential. Cyclic AMP might work by increasing peak Ca^{++} influx into the terminals. A similar increase in Ca^{++} influx has been observed into heart muscle fibers in response to the excitatory transmitter norepinephrine (Tsien, 1977).

Recently Marc Klein, a graduate student in my laboratory, and I have obtained some evidence in support of this scheme. We have found that the presynaptic facilitation is associated with an enhanced activation of the voltage-dependent Ca^{++} conductance turned on by the depolarizing action of the action potential.

The best way to examine the action of cAMP on the terminals would obviously be to record from the terminals themselves. Unfortunately, we have not been able to do this. However, we have found, as has previously been noted by Nicholls and Purves (1970) in the leech, that the terminals are sufficiently close to the cell body that some of the electrical changes in them are reflected by changes in the cell body. In addition, the properties of the Ca^{++} channels of the cell body seem to resemble those of the terminals. This also had previously been suggested by the work of Stinnakre and Tauc (1973).

We therefore set about to examine the changes in the membrane properties of the cell body that accompany the presynaptic facilitation and found two types of change, each of which persists up to an hour. One, which I will not consider here, is a small depolarization, often negligible at the resting level. This slow EPSP is simulated by serotonin and by cyclic AMP, and is enhanced by 3-isobutyl-1-methylxanthine (IBMX), an agent that inhibits the enzyme phosphodiesterase (Klein and Kandel, 1978).

More relevant for the discussion of presynaptic facilitation, we think, is the second type of change that is produced by presynaptic facilitatory input on the voltage-dependent conductances underlying the action potential. The magnitude of the conductance changes produced in the postsynaptic cells by chemical synaptic transmitters has generally been found to be relatively independent of membrane potential. These conductances are quite distinct from the voltage-dependent conductances that underlie the generation of the action potential in the postsynaptic cell. In fact, one major feature that conventionally distinguishes the synaptic conductance channels from those of the action potential is the independence of the synaptic channels from membrane potential (for discussion see Grundfest, 1957; Eccles, 1964).* In contrast to these conventional synaptic actions, the modulatory synaptic actions produced by the facilitatory neurons activate voltage-dependent conductance channels in the presynaptic terminals of the sensory neurons. These channels are turned off at the resting level of membrane potential but are turned on by the voltage swing produced by the action potential.

---

*For interesting exceptions see the studies on the heart reviewed by Noble (1977) and the studies in nerve and muscle cells by Magleby and Stevens (1972), Hartzell, Kuffler, Stickgold, and Yoshikami (1977), and Minota and Koketsu (1977).

A first clue that a voltage-sensitive Ca^{++} conductance might be involved was the finding that stimulation of the connectives, one of the procedures that elicits presynaptic facilitation, often produced a small change in the configuration of the spike. This small change was difficult to study, but the failure to see a larger change was not surprising. As is well known from the work of Katz and Miledi (1969), the Ca^{++} current in the terminals is normally masked by the much larger Na^{+} and K^{+} current. In particular, the K^{+} current overlaps much of the duration of the Ca^{++} current (Figure 33*A*). We therefore incubated the ganglion with tetraethylammonium (TEA), an agent that selectively blocks one of the delayed K^{+} channels. TEA produces a significant prolongation of the spike because it blocks the repolarizing action of the K^{+} current (Figures 33*B* and 34*A* and *B1*). Much of this prolongation is due to the unopposed action of the Ca^{++} current (see also Horn and Miller, 1977). In the presence of TEA, the Ca^{++} current is sufficiently unmasked so that it can itself produce a depolarization of the membrane, which in turn can further prolong and enhance the Ca^{++} conductance increase. Thus, the prolongation in the sensory neuron spike is almost totally prevented by Co^{++}, which blocks the Ca^{++} channels of neurons selectively (Figure 34*B* and Hagiwara, 1973). Moreover, the late phase of the plateau in TEA has the properties of a Nernst calcium electrode. A 10-fold change in Ca^{++} concentration produces a 28 mV change in the amplitude of the late phase of the plateau. In addition, as I will show below, the prolongation produced by TEA occurs in the absence of external Na^{+}.

As Katz and Miledi (1969) pointed out, in the presence of TEA the duration of the spike is very sensitive to slight changes in Ca^{++} current. The prolonged, TEA-treated spike therefore is a good assay for changes in Ca^{++} current, although the assay does not distinguish whether the action is on the Ca^{++} channel directly or whether that channel is affected indirectly by the inactivation of a K^{+} conductance.

Klein and I were now in a position to see fairly directly how the Ca^{++} current is altered by: (1) stimulation of the connectives, (2) incubation with serotonin, the putative mediating transmitter, (3) intracellular injection of cAMP, the possible second messenger, and (4) incubation with IBMX, a phosphodiesterase inhibitor that enhances the action of endogenous cAMP. As is illustrated in Figure 35, all four procedures prolong the TEA-treated spike, further indicating that all of these prolong the Ca^{++} current, directly or indirectly. By contrast, intracellular injection of 5′AMP does not produce broadening.

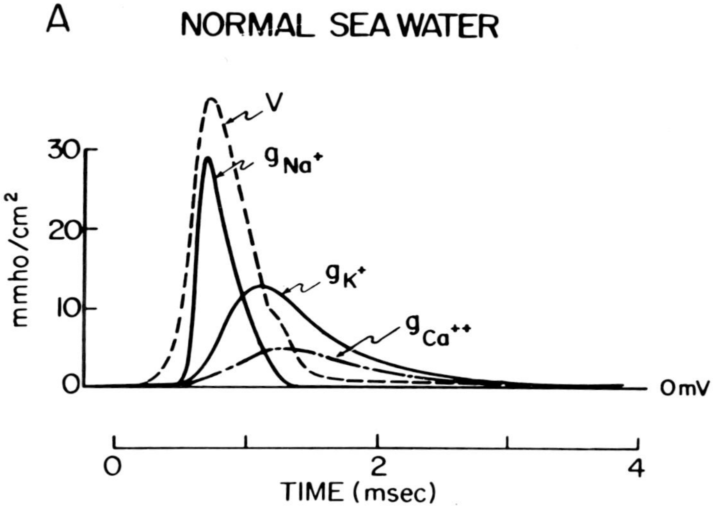

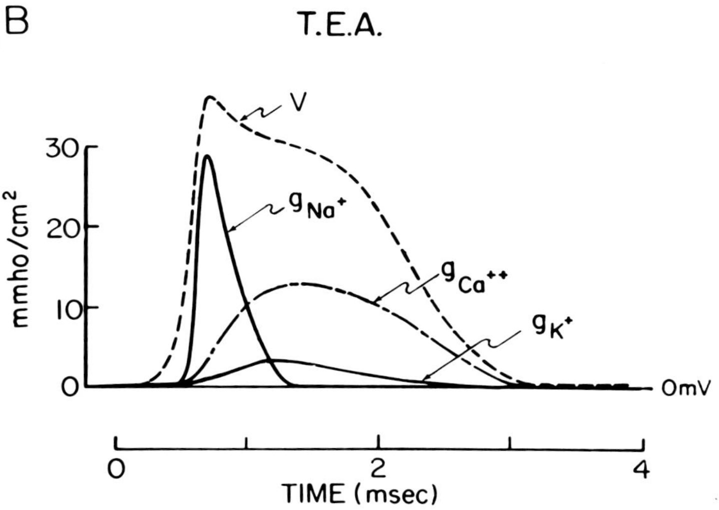

FIGURE 33. Schematic diagram illustrating the action potential (V) and the sodium, potassium, and calcium conductances, g_{Na}, g_K, and g_{Ca}, in normal seawater (Part *A*) and in tetraethylammonium (TEA; Part *B*). These conductance changes give rise to the currents that produce the action potential. TEA blocks g_K, which results in a slowed repolarization which unmasks and enhances the Ca^{++} current.

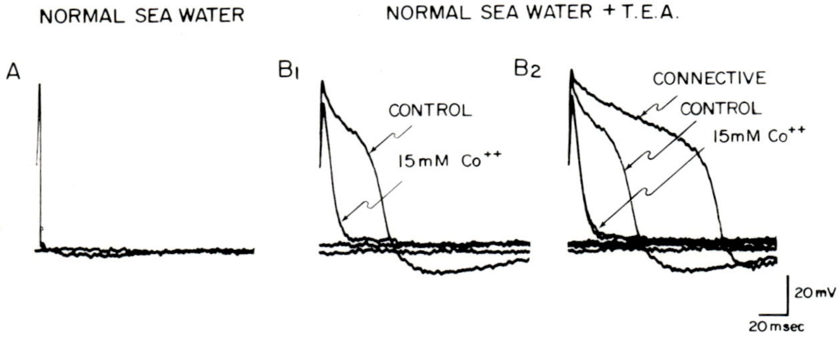

FIGURE 34. The effects of cobalt, a Ca^{++}-channel blocking agent, on the duration of the sensory neuron spike in tetraethylammonim (TEA) and in TEA following connective stimulation.

A: The action potential in normal seawater.

B: The action potential in TEA (which blocks the delayed K^+ channel) before and after connective stimulation. *B1*: The action potential in TEA is greatly prolonged compared to the action potential in seawater. This prolongation is due to an unmasking of and an increase in Ca^{++} conductance and is largely blocked by 15 mM cobalt, which also reduces the overshoot of the action potential. *B2*: Connective stimulation further prolongs the duration of the action potential in TEA; this component of the prolongation also seems to be due to a Ca^{++} conductance, since it is blocked by 15 mM cobalt. (From Klein and Kandel, 1978.)

An independent check on the nature of the current enhanced by these procedures can be obtained by examining the action potential in a Na^+-free solution when the delayed K^+ channels are blocked with TEA. In the absence of Na^+, Ca^{++} is the only likely charge carrier for inward current. With the delayed K^+ blocked by TEA, depolarization of the membrane produces a large, all-or-none Ca^{++} spike, which can be blocked by Co^{++}. The duration of this regenerative Ca^{++} spike is also greatly enhanced by serotonin, and this enhancement is blocked by Co^{++} (Figure 36).

Tomosky-Sykes (unpublished observation) has now also used this assay system to reexamine possible candidates for the transmitter that mediates presynaptic facilitation. He has tested 25 substances, including various amino acids, histamine, octopamine, and dopamine. Of these, only serotonin (and to a lesser degree the related compound 6-HT) produced significant broadening of the action potential, and the broadening was blocked by cinanserin. In addition, Hawkins, Klein, and I (1978*b*)

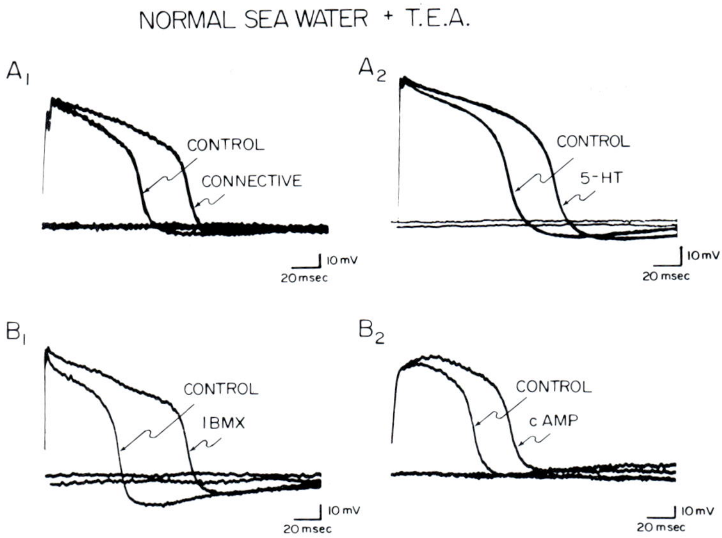

FIGURE 35. Spike broadening may be obtained by intervention at several steps in the proposed mechanism for presynaptic facilitation of the sensory neuron-to-motor neuron synapse.

A1: Connective stimulation. *A2*: Incubation with 10^{-4} M serotonin (5-HT).

B1: Incubation with the phosphodiesterase inhibitor IBMX, which reduces cAMP hydrolysis, leading to increased intracellular levels of cAMP. *B2*: Intracellular injection of cyclic AMP. (From Klein and Kandel, 1978.)

have found that firing a single facilitator neuron also produces broadening of the TEA-treated sensory neuron action potential.

The finding that presynaptic facilitation produces a prolonged activation of a voltage-sensitive Ca^{++} conductance is interesting because it provides a powerful and effective mechanism for controlling synaptic effectiveness. Moreover, this control mechanism might work in both directions and thereby allow synaptic transmission to be depressed as well as enhanced. For example, habituation could produce a progressive decrease in Ca^{++} conductance as a result of the repeated invasion of the terminals by action potentials. To examine this possibility we have begun to look at the action potential of the sensory neurons in TEA during repeated stimulation at rates that produce habituation, and have found that the spike narrows progressively, indicating a progressive decrease in the Ca^{++} current. This decrease in the duration of the

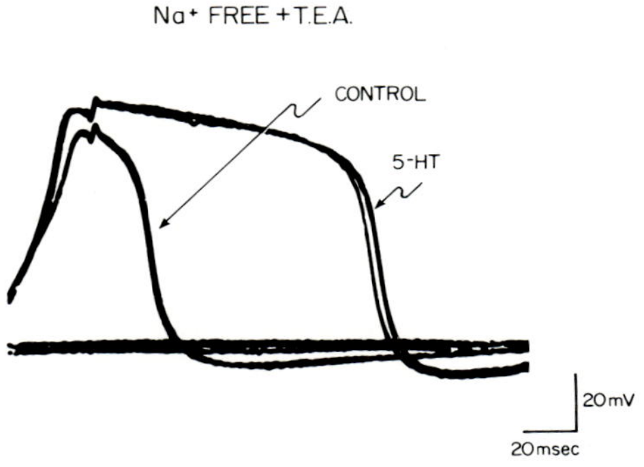

FIGURE 36. Effect of serotonin on a sensory neuron action potential in sodium-free solution containing tetraethylammonium. In sodium-free seawater the blocking of the delayed K^+ channel by TEA makes the calcium component of the action potential large and regenerative. Serotonin produces a slight increase in the rise and in the height of the action potential and a marked increase in the duration of the spike. This spike is mediated by Ca^{++} and is fully blocked by cobalt. Several traces have been superimposed. (From Klein and Kandel, 1978.)

action potential parallels the decrease in the amplitude of the EPSP. Stimulation of the connectives now produces a rapid enhancement of the duration of the action potential and facilitates release (Figure 37).

The diagram of Figure 38 summarizes our speculations about how cyclic AMP might produce its actions. In the resting state the depolarization produced by each action potential opens up a certain number of Ca^{++} channels (the activated Ca^{++} channels are indicated by the black squares in Figure 38). Ca^{++} flows into the terminals, helps to bind the synaptic vesicles to the external membrane, and causes them to release. We posit that sensitization is mediated by serotonergic cells: these increase cAMP, which acts to enhance transmitter release by increasing the Ca^{++} conductance.

It is tempting to think that cyclic AMP might act on the Ca^{++} channels to enhance transmitter release by means of one or more protein kinases—enzymes that phosphorylate proteins. These enzymes have been suggested by Greengard (1976) to be the common molecular site of action for cAMP. Phosphorylation—the covalent insertion of a phosphoryl group to a specific amino acid residue on the protein—changes the charge and configuration of proteins and thereby alters their function in various ways, activating some and inactivating others.

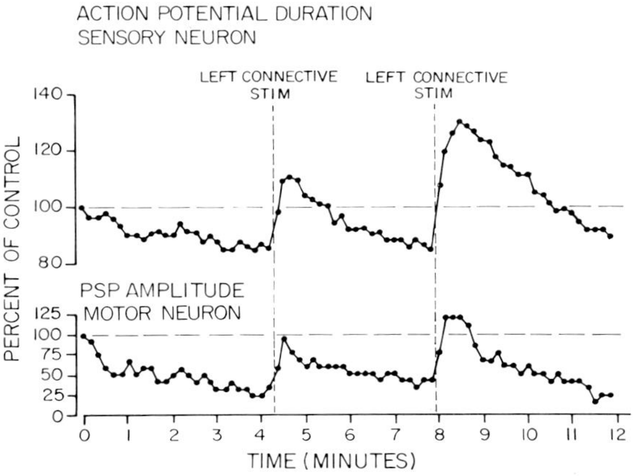

FIGURE 37. Changes in the duration of the action potential of a sensory neuron treated with TEA and the concomitant changes in the EPSP produced by the sensory neuron on the motor neuron L7. Repeated intracellular stimulation of the sensory neuron, simulating habituation training, produces a decrease in the duration of the action potential in the sensory neuron and a concomitant depression in the EPSP in the motor cell. Connective stimulation, simulating sensitization, increases the duration of the action potential in the presynaptic sensory neuron and also increases the amplitude of the EPSP in the motor neuron. The first stimulus to the connective is just above threshold for facilitation; the second stimulus is slightly stronger. (From Klein and Kandel, 1978.)

It could therefore serve as an effective mechanism for long-term regulation. As is the case for other membrane functions, ionic channels are thought to be proteins (Singer and Nicolson, 1972; Benzer and Raftery, 1973; Hille, 1977). One way in which sensitization might work is that the calcium channel protein becomes activated when phosphorylated by a cAMP-dependent protein kinase. It would therefore be interesting to see whether sensitization leads to an increased phosphorylation in the external membrane of the sensory neurons.

Cyclic AMP-dependent protein phosphorylation might affect the Ca^{++} channels directly in one of four ways: (1) by allowing new Ca^{++} channels to be inserted into the membrane, (2) by increasing the conductance of preexisting Ca^{++} channels, (3) by reducing the activation threshold of each Ca^{++} channel, or (4) by increasing the open time of each Ca^{++} channel that is activated.

Alternatively, the cAMP could inactivate the K^+ channel, and the subsequent depolarization would therefore indirectly prolong the action

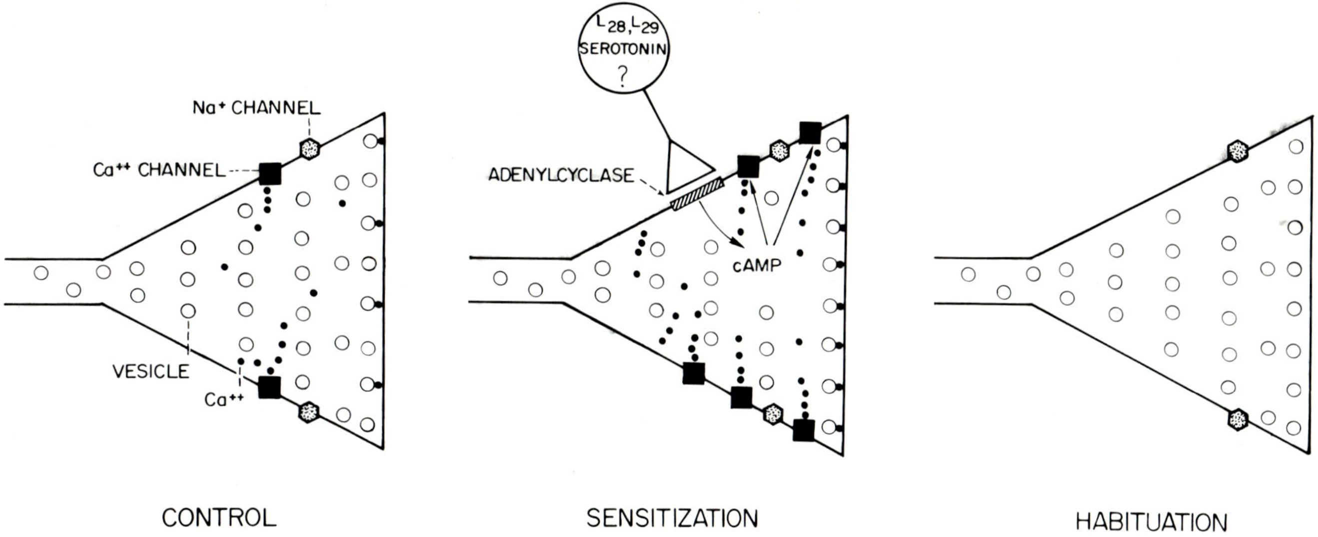

FIGURE 38. A model for sensitization and habituation. In the control state an action potential in the terminal membranes of the sensory neurons opens up a number of Ca^{++} channels (squares) in parallel with the Na^{+} channels of the membrane (hexagons). Sensitization is produced by cells L28 and L29 (and perhaps by some other cells); these cells are thought to be serotonergic. Serotonin acts on an adenylate cyclase in the terminals, which stimulates the synthesis of cAMP. cAMP in turn increases the availability of Ca^{++} channels, leading to an increase in g_{Ca}, a greater influx of Ca^{++} and greater binding of vesicles to release sites, and therefore to an increased probability of release. Repeated action potentials in the terminals accompanying habituation decrease the number of open Ca^{++} channels in the sensory terminal and in the limit may shut them down altogether. The resulting depression in Ca^{++} influx would thereby functionally inactivate the synapse, because Ca^{++} is thought to be required for the vesicles to bind to release sites.

of the Ca^{++} channels. Although it is somewhat unlikely, we cannot exclude for certain the possibility that the residual delayed K^+ current not blocked by TEA is reduced further during sensitization. A perhaps more likely alternative is that another K^+ current is involved. As the data of Figure 34*B1* and *B2* show, in the presence of TEA the hyperpolarizing afterpotential of the spike is enhanced. Preliminary experiments by Klein and myself (1978) suggest that this enhancement may be due to a K^+ conductance that is turned on by the Ca^{++} influx (see Meech and Standen, 1975). Sensitization might therefore operate by reducing the coupling between Ca^{++} influx and the Ca^{++}-dependent K^+ channel. However, we have found that in a high-calcium solution (without TEA) connective stimulation enhances the rate of rise of the ascending limb of the action potential and increases the overshoot. A similar increase in the rate of rise of the ascending limb and in the overshoot is also observed in the Ca^{++} spike studied in a Na^+-free solution with TEA. It is likely that there is a lag between Ca^{++} influx that contributes to the ascending limb of the action potential and the activation of a Ca^{++}-dependent K^+ conductance. Thus, although not conclusive, the effects observed on the early phases of the action potential favor a direct action of the Ca^{++} conductance.

Habituation seems to involve a decrease in the Ca^{++} conductance. It is likely, however, that additional factors are operative, since with prolonged or repeated periods of stimulation the depression in transmitter release is more profound than the decrease in the duration of the action potential would suggest. (A similar dissociation of transmitter depression and depression of Ca^{++} potentials or current occurs at the giant synapse of the squid; see Katz and Miledi, 1969; Llinás, 1977.)

I must emphasize that three critical aspects of this model of sensitization are based on rather indirect evidence. We therefore need to determine more directly: (1) Whether or not the presynaptic facilitating neurons are serotonergic. We have so far provided some pharmacological support for this idea, but we need to dissect out individual facilitatory neurons and show biochemically that they synthesize serotonin. (2) Whether the action of the presynaptic facilitatory cells increases the endogenous level of cyclic AMP in the terminals of the sensory cells. Of the three essential steps necessary to demonstrate cyclic nucleotide participation—simulation by cAMP, enhancement by phosphodiesterase inhibitors, increase of cAMP by the transmitter and by neural stimulation—no single step is conclusive and we are lacking evidence on the last point. Whereas connective stimulation or serotonin will increase cAMP in the whole abdominal ganglion, we do not as yet know that these manipulations increase the level of cAMP in individual sensory neurons. And (3) whether the increase in Ca^{++} conductance produced by the synaptic action is direct or whether the calcium channels stay open because a competing K^{+} conductance is inactivated.

A SEARCH FOR UNIFYING PRINCIPLES UNDERLYING SENSITIZATION AND AROUSAL

The known ubiquity of the Ca^{++}- and cAMP-interdependent mechanisms (Rasmussen, 1975) raises two further questions I would like to consider: (1) Do other forms of arousal use similar mechanisms, and (2) do forms of behavioral modulation other than arousal use analogous mechanisms?

Common Features in the Sensitization of Gill Withdrawal, Feeding, and Heart Rate

In addition to our own group, several others working on *Aplysia* have independently been studying aspects of arousal. These groups are now taking advantage of the animal's limited behavioral repertory, which I have referred to before, to examine the mechanistic interrelationships between the modulation of different responses.

In the course of studying feeding, a complex fixed-action pattern of *Aplysia,* Kupfermann and his colleagues have found that exposure of *Aplysia* to food produces sensitization of the biting response. The latency of successive biting responses decreases and their amplitude increases (Kupfermann, 1974; Susswein, Weiss, and Kupfermann, 1978; Weiss and Kupfermann, 1978). In a parallel study, Dieringer et al. (1978) found that arousal by food stimuli produces a concomitant increase in heart rate.

As I indicated earlier, aspects of the neural circuit controlling feeding are now well understood (Cohen, Weiss, and Kupfermann, 1978; Weiss, Cohen, and Kupfermann, 1978), as are aspects of the neural circuit controlling heart rate (Mayeri et al., 1974; Koester et al., 1974). In each of these circuits, serotonergic cells have been found to have a prominent modulating role. For example, Weiss and Kupfermann (1976) have found that the symmetrical metacerebral cells of the cerebral ganglia seem to be critically involved in mediating sensitization of biting (Weiss, Cohen, and Kupfermann, 1975, 1978). The metacerebral cells are normally silent, but brief exposure of the animal to food activates the cells and causes them to fire. The metacerebral cells in turn influence the muscle directly to increase its strength of contraction by enhancing excitation-contraction coupling. As is the case for the facilitating neurons L28 and L29 in the gill-withdrawal reflex, the action of the metacerebral cells on muscle is modulatory, not mediating. Activity of the metacerebral cells does not produce a detectable synaptic potential or contraction in the muscle. Rather, the cells enhance the contraction produced by the motor neurons that innervate the muscles of the buccal mass, the feeding organ of *Aplysia* (Figure 39*A*).

The metacerebral cells are serotonergic (McCaman and Dewhurst, 1970; Eisenstadt et al., 1973; Gerschenfeld and Paupardin-Tritsch, 1974) and mediate their sensitizing action on the buccal muscle by means of cAMP (Weiss, Schonberg, Cohen, Mandelbaum, and Kupfermann, 1976). Serotonin, applied to the intact muscle or to cell-free homogenates, increases the synthesis of cAMP. Weiss, Schonberg, Mandelbaum, and Kupfermann (1978) have found that firing a single metacerebral cell can cause a marked increase in the synthesis of cAMP in the buccal muscle. In addition, analogs of cAMP simulate the enhanced contraction produced by the metacerebral cells (Figure 39*B*).

A similar mechanism seems to apply to the sensitization of heart rate following the presentation of food. The major heart excitor, RB_{HE} (Mayeri et al., 1974), has been found to account for at least a portion of this heart rate increase (Koester and Weiss, unpublished observation). This cell is serotonergic and produces a prolonged speeding up of heart rate (Figure 40; Mayeri et al., 1974; Liebeswar, Goldman, Koester, and Mayeri, 1975). Heart rate increases produced by direct application of serotonin to the heart are associated with increased levels of cyclic AMP (Mandelbaum, Koester, Weiss, and Schonberg, unpublished observation). This suggests that the heart excitor may exert its motor effect by means of an increase in cyclic AMP.

Comparison of the results in the three systems leads to two conclusions: (1) that on the cellular level, arousal in *Aplysia* is not unitary; it is not mediated by a single cell or by a topographically related group of cells. Rather, different neurons located in different ganglia are involved in modulating the gain of different behavioral responses. And (2) that despite this distributed nature, the various arousal systems share aspects of a common mechanism (Figure 41). It is tempting to speculate that in each case serotonin might act by means of cAMP to control a voltage-dependent Ca^{++} conductance. In the case of gill withdrawal the control is exerted in the presynaptic terminals, where the

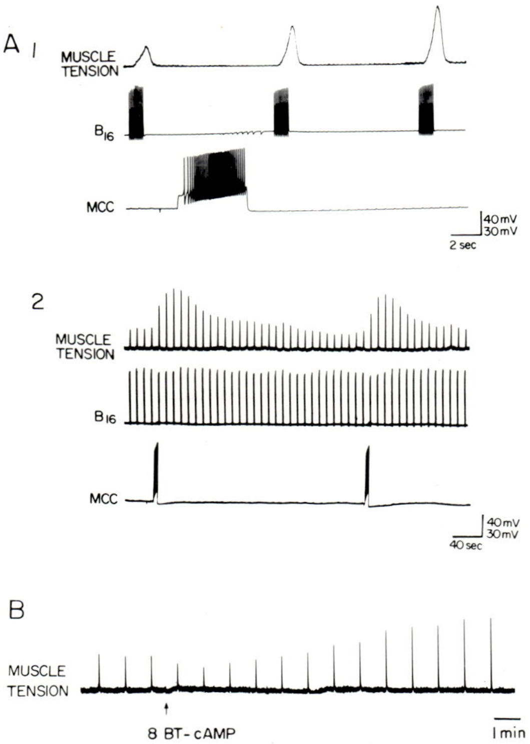

FIGURE 39. Role of the metacerebral cells in the sensitization of biting in *Aplysia*.

A: Potentiation of buccal muscle contraction produced by an identified buccal motor neuron, B_{16}. *A1*: Fast-sweep record showing muscle contractions produced by a fixed number at a constant frequency of motor neuron spikes before and after stimulation of a metacerebral cell (MCC). *A2*: Slow-sweep record showing the time course of MCC potentiation. The MCC was stimulated twice to show the reproducibility of the effect.

B: Potentiation of muscle contraction by an analog of cyclic AMP. Muscle contractions were produced by a constant burst of motor neuron spikes every 10 sec. At the arrow the preparation was perfused with 5×10^{-4} M 8-benzylthio-cAMP. (From Weiss et al., 1976.)

inward Ca^{++} current could enhance transmitter release. In the case of heart rate the action is on the muscle directly, where Ca^{++} could control the strength and frequency of contraction (for earlier and parallel studies on the verebrate heart, see Tsien, 1977). In the case of biting the

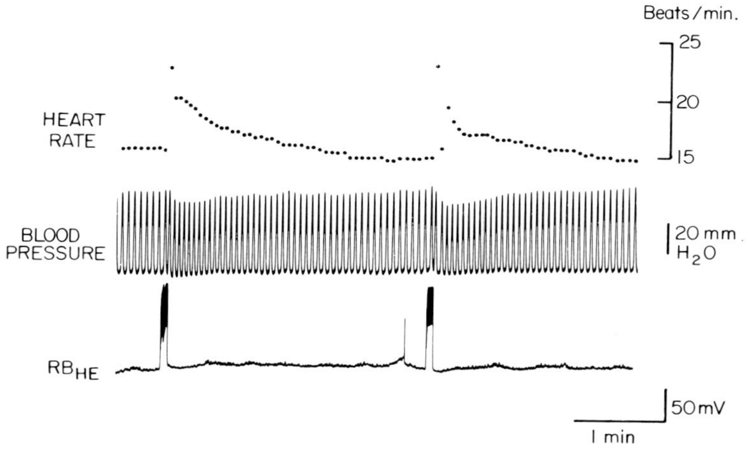

FIGURE 40. Prolonged increase in heart rate produced by the serotonergic heart excitor RB_{HE}. Long-lasting increase in heart rate caused by a 3-sec burst of action potentials elicited in RB_{HE} by intracellular depolarization. (From Mayeri et al., 1974.)

control is exerted at three points: (1) on the muscle, (2) on the motor neuron terminals, and (3) centrally, in an as yet unspecified manner, on the central program and by depolarizing the motor neurons. Some of these actions again might be produced by cAMP-mediated changes in Ca^{++} conductance.

A corollary to the finding that serotonin is involved in several instances of arousal in *Aplysia* is that specific transmitters may have specific functions in mediating aspects of arousal or types of learning because of their ability to act on specific biochemical processes in the postsynaptic cell that are capable of regulating the long-term activity of the neuron (Cedar et al., 1972).

Other Types of Response Modulation

Prolonged synaptic modulation of voltage-dependent ionic channels could provide a common means for long-term modulation of neuronal activity. This type of modulation is highly efficient in that it is called into play only when signaling occurs. The channel is activated only when brought into appropriate operating range by the depolarizing action of a synaptic potential or an action potential. The type of neuronal modula-

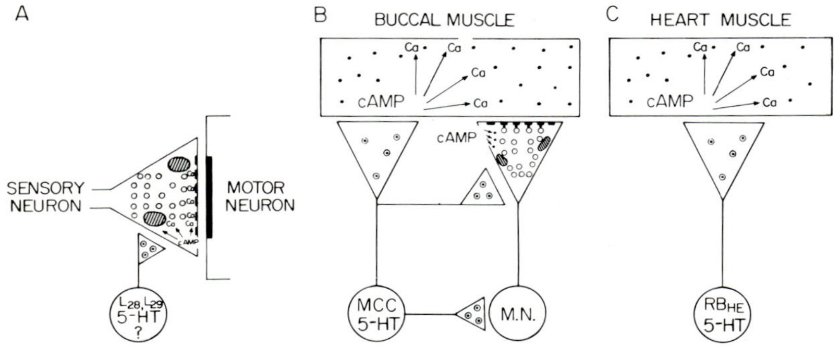

FIGURE 41. Schematic diagram illustrating common features in the arousal (sensitizing) of gill withdrawal, of heart rate, and of feeding. Cyclic AMP increase is indicated by arrows; the dots are free Ca^{++} ions. 5-HT-containing vesicles are schematized as dense-core vesicles.

tion achieved would depend upon the *nature* of the channel and its *location.* Thus, modulation of Ca^{++} channels in the terminal of a neuron would alter transmitter release. It could enhance and, in principle, also depress release. Indeed, there is now evidence that synaptic modulation of a voltage-dependent Ca^{++} channel accounts for a prolonged type of presynaptic inhibition that seems to operate in pain pathways of the vertebrate brain. Some of the dorsal root ganglion cells with small diameter axons that may serve as sensory neurons in the pathways for pain contain substance P. These neurons release substance P on stimulation, a finding that suggests that this peptide is involved in synaptic transmission (Otsuka and Konishi, 1976; Hökfelt, Kellerth, Nilsson, and Pernow, 1977). Jessell and Iversen (1977) have recently found that, in the trigeminal nucleus of the rat, opiate analgesics and the endogenous opioid peptide enkephalin suppress the release of substance P from the sensory neurons. Jessell and Iversen suggest that the enkephalins may act directly on the sensory neurons.

This hypothesis has received direct support by the experiments of Mudge and Fischbach (personal communication). They found that substance P is present in dissociated sensory neurons of chick embryo grown in tissue culture in the absence of other types of neuronal or non-neuronal cells. These cells release substance P when depolarized with a high K^{+} medium, and this release is inhibited, presynaptically, by enkephalin. The presynaptic inhibition is associated with a decrease in the duration of the action potential. When enkephalin is applied directly to the cell body, the action potential is reduced in duration. In an earlier study, Dichter and Fischbach (1977) found that the cell bodies of these embryonic chick neurons resemble those of *Aplysia* in having both a Ca^{++} and a Na^{+} component to their spike. Thus the reduced spike duration is consistent with an action of enkephalin on a voltage-sensitive Ca^{++} conductance leading to a reduced Ca^{++} influx into the neuron (Figure 42*A*). It will be of interest to see whether this action also involves cyclic nucleotides.

A voltage-dependent conductance that is modulated by synaptic input need not be located in the terminal of the neuron and need not participate directly in controlling transmitter release. The synaptic modulation of a voltage-dependent Ca^{++}, Na^{+}, or K^{+} channel in the cell body or axon (the trigger zone or spike-generating part of the neuron) could alter the threshold of a neuron, turn a silent cell into an autoactive one, or change the pattern of firing of an autoactive cell. As Smith et al. (1975) have first shown, peptide hormones can modulate the endogenous firing patterns of neurons by acting on voltage-sensitive ionic conductances (Figure 42*B*). The earlier findings by Smith, Barker, and Gainer on the modulation of autoactive cells by peptides and the more recent findings by Klein and myself on the modulation of transmitter release by biogenic amines suggest that what appear on the surface to be very different mechanisms for neuronal modulation initiated in different ways—by a neurohormone released at a distance in one case, and by a neural transmitter released at close range in another—might share certain formal properties, including the modulation of a voltage-sensitive conductance by a chemical mediator. Furthermore, the distinction between neuronal transmitters and neuronal hormones may depend less on the chemical nature of the mediator released by the presynaptic neuron than on the nature of the receptor and the subsequent biochemical steps initiated by the mediator in the postsynaptic cell.

Response Modulation and Learning

In discussing the possible generality of voltage-dependent synaptic conductance for response modulation, I am not suggesting that learning or behavioral modulation partakes only of this class of mechanisms. As I have stressed earlier, behavior can be modified in many ways and is likely to involve a variety of mechanisms.

I have extended the discussion of voltage-sensitive conductances and sensitization to include certain other types of response modulation

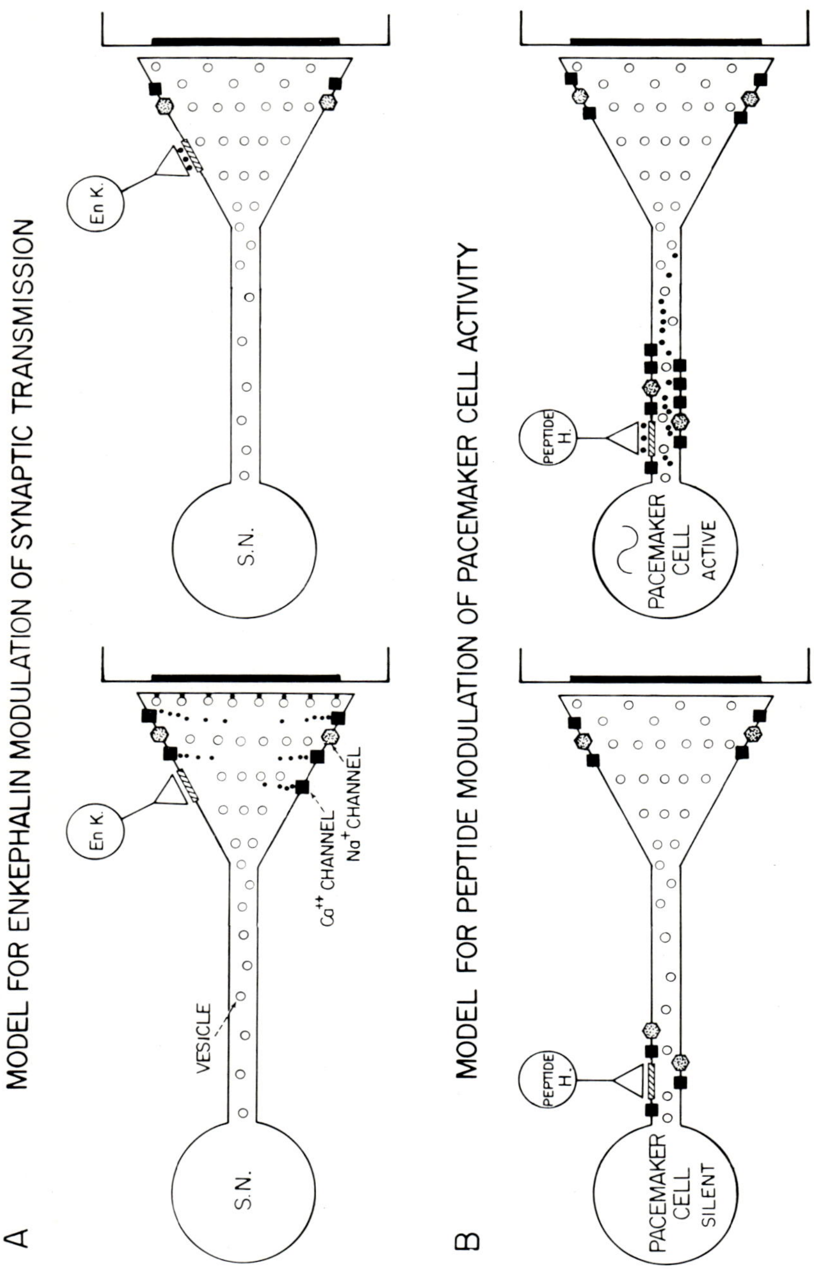

FIGURE 42. See caption on following page.

primarily to indicate that some of the mechanisms underlying the various forms of learning are likely to be used for other purposes in nerve cells and even in non-neuronal cells. For example, the model that I propose here for sensitization has certain specific parallels to that emerging from the earlier analysis of the action of norepinephrine on the pacemaker cells of the vertebrate heart (Tsien, 1973). In more general terms this mechanism resembles that of a particular class of hormones, the peptides and the biogenic amines. These hormones do not enter the target cell but only act on its surface as a trigger or first messenger. This action on the external surface of the membrane leads to an intracellular increase in a second messenger, cyclic AMP, which in turn regulates a family of responses by acting on a variety of molecular regulatory processes. The similarities between the actions of peptide hormones on non-neuronal target cells and of certain biogenic amine transmitters on nerve cells suggest that neurons can use cellular regulatory mechanisms found in non-neuronal cells in novel ways to achieve storage of information necessary for the readout of memory.

I have therefore extended the discussion of learning to neuronal regulation in the same spirit that I proposed, at the beginning of this lecture, the broader rather than the narrow definition of learning, because I believe that learning is shrouded in surplus meaning that hinders rather than facilitates its study. One of the initial tasks confronting the study of learning—which the cellular approach has been helpful in accomplishing—is to demystify the problem and to position it within the boundaries of cell biology.

FIGURE 42. Schematic hypothesis illustrating the possible role of a chemically activated voltage-dependent Ca^{++} conductance in regulating neuronal activity.

A: Prolonged presynaptic inhibition mediated by enkephalin on the terminals of a sensory neuron in the vertebrate pain pathway. These terminals release substance P. Enkephalin seems to produce its inhibitory modulatory action by reducing g_{Ca}. (Based on findings of Mudge and Fischbach, unpublished observation.)

B: Peptide modulation of endogenous rhythm in cells capable of autoactivity. Peptides such as ADH can affect the endogenous rhythms of cells by acting on voltage-sensitive conductances. For simplicity, only a Ca^{++} channel is illustrated as being activated. (Based on findings of Barker and Smith, 1977.)

PERSPECTIVES

In conclusion, I would like to consider two ideas—one practical, the other philosophical.

From a practical point of view, having looked at sensitization, we should not find it too difficult to consider the next step in the hierarchy of learning—associative learning. Several simple gastropod preparations have already been shown capable of some form of associative learning: *Limax* (Gelperin and Forsythe, 1976), *Pleurobranchaea* (Mpitsos and Davis, 1973; Mpitsos and Collins, 1975), and *Hermissenda* (Crow and Alkon, 1977). We can therefore expect a fruitful attack on this problem in the near future. It will be particularly interesting, as these analyses progress, to see what relation the mechanisms of classical conditioning will have to those of sensitization.

From a philosophical point of view, I would like to emphasize that the cellular study of behavior has proven, as expected, to be a two-way street—as interesting from a cell-biological point of view as it is from a behavioral one. Indeed, the growing partnership between the two once-independent fields of study is beginning to lead to important new syntheses in several areas. From the cellular point of view, interesting new cellular properties are revealed time and again during studies of learning—presynaptic facilitation, synapse inactivation, modulation by the action potential of voltage-sensitive synaptic conductances. These properties have come to light because a particular cell located in a particular neural circuit was examined with a particular form of learning in mind. In addition to exposing one to a new range of mechanisms, the study of learning infuses the study of cellular mechanisms with a new fascination—a fascination stemming from the knowledge that one is working on the fringes of important mental processes. On the other hand, from a behavioral point of view, cellular approaches have made modest headway, both conceptually and substantively, in beginning to answer some unsolved questions of learning psychology: How does habituation relate to dishabituation? How does short-term memory relate to long-term memory?

Having stated that headway has been made, I hasten to emphasize that we are still at the very beginning of understanding the mechanisms of behavior and learning. We have so far analyzed only simple non-associative forms of learning in simple behavior. We still do not understand the mechanisms of the various forms of associative learning in simple behavior, nor do we understand any mechanisms of learning involving complex behavior. Clearly the remainder of the road is long. Nonetheless, those of us working in the field find it encouraging to think that we are on the right road and that by taking several large reductionist steps backward, cell biologists have begun to redirect the study of learning to new paths, to where the footing is surer. As a result one can now begin to take some small but confident steps forward.

REFERENCES

Advokat, C. (1978). Modulation of simple defensive reflexes in *Aplysia californica* by appetitive stimuli, in preparation.

Agranoff, B. W. (1967). Agents that block memory, pp. 756–764 in *The Neurosciences: A Study Program,* Quarton, G. C., T. Melnechuk, and F. O. Schmitt, eds. Rockefeller University Press, New York.

Arvanitaki, A. and N. Chalazonitis (1958). Configurations modales de l'activité, propres à differénts neurones d'un même centre, *J. Physiol. (Paris)* **50**:122–125.

Arvanitaki, A. and S. H. Tchou (1942). Les lois de la croissance relative individuelle des cellules nerveuses chez l'Aplysie, *Bull. Histol. Appl. Physiol. Pathol.* **19:**244–256.

Bailey, C. H. and E. B. Thompson (1978). Cup-shaped synapses in *Aplysia,* in preparation.

Bailey, C. H., E. B. Thompson, V. F. Castellucci, and E. R. Kandel (1976). Fine structure of synapses of identified sensory cells which mediate the gill-withdrawal reflex in *Aplysia californica*, *Soc. Neurosci., Abstr.* Vol. 2, p. 446.

Bailey, C. H., E. B. Thompson, V. F. Castellucci, and E. R. Kandel (1978). Fine structure of synapses of identified sensory cells which mediate the gill-withdrawal reflex in *Aplysia californica, Soc. Neurosci. Abstr.,* Vol. 2, p. 446.

Barker, J. L. and T. G. Smith, Jr. (1977). Peptides as neurohormones, pp. 340–373 in *Society for Neuroscience Symposia,* Vol. II, Cowan, W. M. and J. A. Ferrendelli, eds. Society for Neuroscience, Bethesda, Md.

Barondes, S. H. (1970). Multiple steps in the biology of memory, pp. 272–278 in *The Neurosciences: Second Study Program,* Schmitt, F. O., G. C. Quarton, T. Melnechuk, and G. Adelman, eds. Rockefeller University Press, New York.

Benzer, T. I. and M. A. Raftery (1973). Solubilization and partial characterization of the tetrodotoxin binding component from nerve axons, *Biochem. Biophys. Res. Commun.* **51**:939–944.

Berry, M. S. and G. A. Cottrell (1975). Excitatory, inhibitory and biphasic synaptic potentials mediated by an identified dopamine-containing neurone, *J. Physiol. (London)* **244**:589–612.

Brunelli, M., V. Castellucci, and E. R. Kandel (1976). Synaptic facilitation and behavioral sensitization in *Aplysia:* possible role of serotonin and cyclic AMP, *Science (Wash., D.C.)* **194**:1178–1181.

Byrne, J., V. Castellucci, and E. R. Kandel (1974). Receptive fields and response properties of mechanoreceptor neurons innervating siphon skin and mantle shelf in *Aplysia, J. Neurophysiol.* **37**:1041–1064.

Byrne, J. H., V. F. Castellucci, and E. R. Kandel (1978). Contribution of individual mechanoreceptor sensory neurons to defensive gill-withdrawal reflex in *Aplysia, J. Neurophysiol.* **41**:418-431.

Callec, J. J., J. C. Guillet, Y. Pichon, and J. Boistel (1971). Further studies on synaptic transmission in insects. II. Relations between sensory information and its synaptic integration at the level of a single giant axon in the cockroach, *J. Exp. Biol.* **55**:123-149.

Carew, T. J., V. F. Castellucci, and E. R. Kandel (1971). An analysis of dishabituation and sensitization of the gill-withdrawal reflex in *Aplysia, Int. J. Neurosci.* **2**:79-98.

Carew, T. J. and E. R. Kandel (1977). Inking in *Aplysia californica.* I. Neural circuit of an all-or-none behavioral response, *J. Neurophysiol.* **40**:692-707.

Carew, T. J., H. M. Pinsker, and E. R. Kandel (1972). Long-term habituation of a defensive withdrawal reflex in *Aplysia, Science (Wash., D.C.)* **175**:451-454.

Carew, T. J., H. Pinsker, K. Rubinson, and E. R. Kandel (1974). Physiological and biochemical properties of neuromuscular transmission between identified motoneurons and gill muscle in *Aplysia, J. Neurophysiol.* **37**:1020-1040.

Castellucci, V. F., T. J. Carew, and E. R. Kandel (1978). Cellular analysis of long-term habituation in the gill-withdrawal reflex of *Aplysia californica, Science* (Wash., D.C.), in press.

Castellucci, V. F. and E. R. Kandel (1974). A quantal analysis of the synaptic depression underlying habituation of the gill-withdrawal reflex in *Aplysia, Proc. Natl. Acad. Sci. U.S.A.* **71**:5004-5008.

Castellucci, V. and E. R. Kandel (1976). Presynaptic facilitation as a mechanism for behavioral sensitization in *Aplysia, Science (Wash., D.C.)* **194**:1176-1178.

Castellucci, V., H. Pinsker, I. Kupfermann, and E. R. Kandel (1970). Neuronal mechanisms of habituation and dishabituation of the gill-withdrawal reflex in *Aplysia, Science (Wash., D.C.)* **167**:1745-1748.

Cedar, H., E. R. Kandel, and J. H. Schwartz (1972). Cyclic adenosine monophosphate in the nervous system of *Aplysia californica.* I. Increased synthesis in response to synaptic stimulation, *J. Gen. Physiol.* **60**:558-569.

Cedar, H. and J. H. Schwartz (1972). Cyclic adenosine monophosphate in the nervous system of *Aplysia californica.* II. Effect of serotonin and dopamine, *J. Gen. Physiol.* **60**:570-587.

Cohen, J. L., K. R. Weiss, and I. Kupfermann (1978). Motor control of buccal muscles in *Aplysia, J. Neurophysiol.* **41**:157-180.

Connor, J. A. and C. F. Stevens (1971*a*). Voltage clamp studies of a transient outward membrane current in gastropod neural somata, *J. Physiol.* (*London*) **213**:21-30.

Connor, J. A. and C. F. Stevens (1971*b*). Prediction of repetitive firing behaviour from voltage clamp data on an isolated neurone soma, *J. Physiol.* (*London*) **213**:31-53.

Crow, T. and D. Alkon (1977). Acquisition and retention of a long-term behavioral change in *Hermissenda crassicornis, Soc. Neurosci. Abstr.*, Vol. 3, p. 174.

Davis, W. J. and R. Gillette (1978). Neural correlate of behavioral plasticity in command neurons of *Pleurobranchaea, Science* (*Wash., D.C.*) **199**:801-804.

Davis, W. J., G. J. Mpitsos, and J. M. Pinneo (1974). The behavioral hierarchy of the mollusk *Pleurobranchaea.* I. The dominant position of the feeding behavior, *J. Comp. Physiol.* **90**:207–224.

del Castillo, J. and B. Katz (1954*a*). The effect of magnesium on the activity of motor nerve endings, *J. Physiol. (London)* **124**:553–559.

del Castillo, J. and B. Katz (1954*b*). Quantal components of the end-plate potential, *J. Physiol. (London)* **124**:560–573.

del Castillo, J. and B. Katz (1954*c*). Statistical factors involved in neuromuscular facilitation and depression, *J. Physiol. (London)* **124**:574–585.

Dichter, M. A. and G. D. Fischbach (1977). The action potential of chick dorsal root ganglion neurones maintained in cell culture, *J. Physiol. (London)* **267**:281–298.

Dieringer, N., J. Koester, and K. R. Weiss (1978). Adaptive changes in heart rate of *Aplysia californica, J. Comp. Physiol.* **123**:11–21.

Eales, N. B. (1921). *Aplysia.* Liverpool Marine Biology Committee, Memoir 24, *Proc. Trans. Liverpool Biol. Soc.* **35**:183–266.

Eccles, J. C. (1964). *The Physiology of Synapses.* Springer-Verlag, Berlin.

Eckert, R. and H. D. Lux (1976). A voltage sensitive persistent calcium conductance in neuronal somata of *Helix, J. Physiol.* (*London*) **254**:129–151.

Eisenstadt, M., J. E. Goldman, E. R. Kandel, H. Koike, J. Koester, and J. H. Schwartz (1973). Intrasomatic injection of radioactive precursors for studying transmitter synthesis in identified neurons of *Aplysia californica, Proc. Natl. Acad. Sci. U.S.A.* **70**:3371–3375.

Frazier, W. T., E. R. Kandel, I. Kupfermann, R. Waziri, and R. E. Coggeshall (1967). Morphological and functional properties of identified neurons in the abdominal ganglion of *Aplysia californica, J. Neurophysiol.* **30**:1288–1351.

Frisch, K. von (1950). *Bees: Their Vision, Chemical Senses, and Language.* Cornell University Press, Ithaca, N.Y.

Gardner, D. and E. R. Kandel (1972). Diphasic postsynaptic potential: a chemical synapse capable of mediating conjoint excitation and inhibition, *Science* (*Wash., D.C.*) **176**:675–678.

Geffen, L. B. and B. G. Livett (1971). Synaptic vesicles in sympathetic neurons, *Physiol. Rev.* **51**:98–157.

Gelperin, A. (1975). Rapid food-aversion learning by a terrestrial mollusk, *Science* (*Wash., D.C.*) **189**:567–570.

Gelperin, A. and D. Forsythe (1976). Neuroethological studies of learning in mollusks, pp. 239–246 in *Simpler Networks and Behavior,* Fentress, J. C., ed. Sinauer Assoc., Sunderland, Mass.

Gerschenfeld, H. M. and D. Paupardin-Tritsch (1974). On the transmitter function of 5-hydroxytryptamine at excitatory and inhibitory monosynaptic junctions, *J. Physiol.* (*London*) **243**:457–481.

Giller, E., Jr. and J. H. Schwartz (1968). Choline acetyltransferase: regional distribution in the abdominal ganglion of *Aplysia, Science* (*Wash., D.C.*) **161**:908–911.

Goldschmidt, R. (1907). Einiges vom feineren Bau des Nervensystems, *Verh. Dtsch. Zool. Ges.,* pp. 130–132.

Greengard, P. (1976). Possible role for cyclic nucleotides and phosphorylated membrane proteins in postsynaptic actions of neurotransmitters, *Nature (London)* **260**:101–108.

Groves, P. M. and R. F. Thompson (1970). Habituation: a dual-process theory, *Psychol. Rev.* **77**:419–450.

Grundfest, H. (1957). Electrical inexcitability of synapses and some consequences in the central nervous system, *Physiol. Rev.* **37**:337–361.

Hagiwara, S. (1973). Ca spike, *Adv. Biophys.* **4**:71-102.

Harlow, H. F. (1958). The evolution of learning, pp. 269–290 in *Behavior and Evolution,* Roe, A. and G. G. Simpson, eds. Yale University Press, New Haven.

Harris, J. D. (1943). Habituatory response decrement in the intact organism, *Psychol. Bull.* **40**:385-422.

Hartzell, H. C., S. W. Kuffler, R. Stickgold, and D. Yoshikami (1977). Synaptic excitation and inhibition resulting from direct action of acetylcholine on two types of chemoreceptors on individual amphibian parasympathetic neurones, *J. Physiol. (London)* **271**:817–846.

Hartzell, H. C., S. W. Kuffler, and D. Yoshikami (1976). The number of acetylcholine molecules in a quantum and the interaction between quanta at the subsynaptic membrane of the skeletal neuromuscular synapse, *Cold Spring Harbor Symp. Quant. Biol.* **40**:175–186.

Hawkins, R. D. (1978). Natural stimuli to the head activate identified neurons mediating presynaptic facilitation in *Aplysia,* in preparation.

Hawkins, R. D. and C. Advokat (1978). The effects of behavioral state on the gill-withdrawal reflex in *Aplysia californica,* pp. 16–32 in *Society for Neuroscience Symposia,* Vol. III, Ferrendelli, J. A., ed. Society for Neuroscience, Bethesda, Md.

Hawkins, R., V. Castellucci, and E. R. Kandel (1976). Identification of individual neurons mediating the heterosynaptic facilitation underlying behavioral sensitization in *Aplysia, Soc. Neurosci. Abstr.,* Vol. 2, p. 325.

Hawkins, R. D., V. F. Castellucci, and E. R. Kandel (1978*a*). Identified neurons mediating the presynaptic facilitation underlying behavioral sensitization in *Aplysia,* in preparation.

Hawkins, R., M. Klein, and E. R. Kandel (1978*b*). Single identified neurons produce presynaptic facilitation of the gill-withdrawal reflex by increasing Ca^{++} conductance in sensory neurons of *Aplysia, Soc. Neurosci. Abstr.,* Vol. 4, p. 196.

Hebb, D. O. (1949). *The Organization of Behavior: A Neuropsychological Theory.* John Wiley & Sons, New York.

Hening, W., T. Carew, and E. Kandel (1977). Motor program for locomotion in *Aplysia* does not require peripheral feedback, *Soc. Neurosci. Abstr.,* Vol. 3, p. 382.

Hilgard, E. R. and D. G. Marquis (1940). *Conditioning and Learning.* Appleton-Century, New York.

Hille, B. (1977). Ionic basis of resting and action potentials, pp. 99–136 in "Cellular Biology of Neurons" (Vol. I, Sect. 1, *Handbook of Physiology: The Nervous System*), Kandel, E. R., ed. Williams and Wilkins, Baltimore.

Hökfelt, T., J.-O. Kellerth, G. Nilsson, and B. Pernow (1977). Experimental immunohistochemical studies on the localization and distribution of substance P in cat primary sensory neurons, *Brain Res.* **100**:235–252.

Horn, R. and J. J. Miller (1977). A prolonged, voltage-dependent calcium permeability revealed by tetraethylammonium in the soma and axon of *Aplysia* giant neuron, *J. Neurobiol.* **8**:399–415.

Horridge, G. A. (1964). The electrophysiological approach to learning in isolatable ganglia, *Anim. Behav. Suppl.* **1**:163–182.

Humphrey, G. (1933). *The Nature of Learning in Its Relation to the Living System.* Harcourt, Brace, New York.

Jessell, T. M. and L. L. Iversen (1977). Opiate analgesics inhibit substance P release from rat trigeminal nucleus, *Nature (London)* **268**:549–551.

Kandel, E. R. (1976). *Cellular Basis of Behavior: An Introduction to Behavioral Neurobiology.* W. H. Freeman, San Francisco.

Kandel, E. R. (1979). *Behavioral Biology of* Aplysia: *A Contribution to the Comparative Study of Opisthobranch Molluscs.* W. H. Freeman, San Francisco.

Kandel, E. R., M. Brunelli, J. Byrne, and V. Castellucci (1976*a*). A common presynaptic locus for the synaptic changes underlying short-term habituation and sensitization of the gill-withdrawal reflex in *Aplysia, Cold Spring Harbor Symp. Quant. Biol.* **40**:465–482.

Kandel, E. R., T. J. Carew, and J. Koester (1976*b*). Some rules relating the biophysical properties of neurons and their patterns of interconnections to behavior, pp. 187–215 in *Electrobiology of Nerve, Synapse, and Muscle,* Reuben, J. P., D. P. Purpura, M. V. L. Bennett, and E. R. Kandel, eds. Raven Press, New York.

Kandel, E. R., W. T. Frazier, R. Waziri, and R. E. Coggeshall (1967). Direct and common connections among identified neurons in *Aplysia, J. Neurophysiol.* **30**:1352–1376.

Kandel, E. R. and W. A. Spencer (1961). Electrophysiology of hippocampal neurons. II. After-potentials and repetitive firing, *J. Neurophysiol.* **24**:243–259.

Kandel, E. R., W. A. Spencer, and F. J. Brinley, Jr. (1961). Electrophysiology of hippocampal neurons. I. Sequential invasion and synaptic organization, *J. Neurophysiol.* **24**:225–242.

Kandel, E. R. and L. Tauc (1964). Mechanism of prolonged heterosynaptic facilitation, *Nature (London)* **202**:145–147.

Kandel, E. R. and L. Tauc (1965). Heterosynaptic facilitation in neurones of the abdominal ganglion of *Aplysia depilans, J. Physiol. (London)* **181**:1–27.

Katz, B. and R. Miledi (1969). Tetrodotoxin-resistant electric activity in presynaptic terminals, *J. Physiol. (London)* **203**:459–487.

Kehoe, J. S. (1969). Single presynaptic neurone mediates a two component postsynaptic inhibition, *Nature (London)* **221**:866–868.

Kehoe, J. S. (1972). The physiological role of three acetylcholine receptors in synaptic transmission in *Aplysia, J. Physiol. (London)* **225**:147–172.

Klein, M. and E. R. Kandel (1978). Presynaptic modulation of voltage dependent Ca^{++} current: mechanism for behavioral sensitization in *Aplysia californica, Proc. Natl. Acad. Sci. U.S.A.* **75**:3512–3516.

Koester, J. and E. R. Kandel (1977). Further identification of neurons in the abdominal ganglion of *Aplysia* using behavioral criteria, *Brain Res.* **121**:1–20.

Koester, J., E. Mayeri, G. Liebeswar, and E. R. Kandel (1974). Neural control of circulation in *Aplysia*. II. Interneurons, *J. Neurophysiol.* **37**:476–496.

Konorski, J. (1948). *Conditioned Reflexes and Neuron Organization.* Cambridge University Press, Cambridge, England. (Reprinted 1968.)

Krasne, F. B. (1969). Excitation and habituation of the crayfish escape reflex: the depolarizing response in lateral giant fibres of the isolated abdomen, *J. Exp. Biol.* **50**:29–46.

Krasne, F. B. (1978). Extrinsic control of intrinsic neuronal plasticity: an hypothesis from work on simple systems, *Brain Res.* **140**:197-215.

Krasne, F. B. and J. S. Bryan (1973). Habituation: regulation through presynaptic inhibition, *Science (Wash., D.C.)* **182**: 590–592.

Kupfermann, I. (1974). Feeding behavior in *Aplysia*: a simple system for the study of motivation, *Behav. Biol.* **10**:1–26.

Kupfermann, I., T. J. Carew, and E. R. Kandel (1974). Local, reflex, and central commands controlling gill and siphon movements in *Aplysia, J. Neurophysiol.* **37**:996–1019.

Kupfermann, I., V. Castellucci, H. Pinsker, and E. Kandel (1970). Neuronal correlates of habituation and dishabituation of the gill-withdrawal reflex in *Aplysia, Science (Wash., D.C.)* **167**:1743–1745.

Kupfermann, I. and E. R. Kandel (1969). Neuronal controls of a behavioral response mediated by the abdominal ganglion of *Aplysia, Science (Wash., D.C.)* **164**:847–850.

Kupfermann, I. and E. R. Kandel (1970). Electrophysiological properties and functional interconnections of two symmetrical neurosecretory clusters (bag cells) in abdominal ganglion of *Aplysia, J. Neurophysiol.* **33**:865–876.

Kupfermann, I., H. Pinsker, V. Castellucci, and E. R. Kandel (1971). Central and peripheral control of gill movements in *Aplysia, Science (Wash., D.C.)* **174**:1252–1256.

Lagercrantz, H. (1976). On the composition and function of large dense cored vesicles in sympathetic nerves, *Neuroscience* **1**:81–92.

Levitan, I. B. and S. H. Barondes (1974). Octopamine- and serotonin-stimulated phosphorylation of specific protein in the abdominal ganglion of *Aplysia californica, Proc. Natl. Acad. Sci. U.S.A.* **71**:1145–1148.

Liebeswar, G., J. E. Goldman, J. Koester, and E. Mayeri (1975). Neural control of circulation in *Aplysia.* III. Neurotransmitters, *J. Neurophysiol.* **38**:767-779.

Llinás, R. (1977). Calcium and transmitter release in squid synapse, pp. 139–160 in *Society for Neuroscience Symposia*, Vol. II, Cowan, W. M. and J. A. Ferrendelli, eds. Society for Neuroscience, Bethesda, Md.

Lorenz, K. (1965). *Evolution and Modification of Behavior.* University of Chicago Press, Chicago.

Magleby, K. L. and C. F. Stevens (1972). The effect of voltage on the time course of end-plate currents, *J. Physiol. (London)* **223**:151-171.

Mayeri, E., J. Koester, I. Kupfermann, G. Liebeswar, and E. R. Kandel (1974). Neural control of circulation in *Aplysia.* I. Motoneurons, *J. Neurophysiol.* **37**: 458–475.

McCaman, R. E. and S. A. Dewhurst (1970). Choline acetyltransferase in individual neurons of *Aplysia californica, J. Neurochem.* **17**:1421–1426.

McCaman, R. E. and D. G. McKenna (1978). Monosynaptic connections between histamine-containing neurons and their various follower cells, *Brain Res.* **141**:165–171.

Meech, R. W. and N. B. Standen (1975). Potassium activation in *Helix aspersa* neurones under voltage clamp: a component mediated by calcium influx, *J. Physiol. (London)* **249**:211-239.

Meyer, D. R. and M. S. Beattie (1977). Some properties of substrates of memory, pp. 145–162 in *Neuropeptide Influences on the Brain and Behavior*, Miller, L. H., C. A. Sandman, and A. J. Kastin, eds. Raven Press, New York.

Minota, S. and K. Koketsu (1977). Effects of adrenaline on the action potential of sympathetic ganglion cells in bullfrogs, *Jpn. J. Physiol.* **27**:353–366.

Mpitsos, G. J. and S. D. Collins (1975). Learning: rapid aversive conditioning in the gastropod mollusk *Pleurobranchaea, Science (Wash., D.C.)* **188**: 954–957.

Mpitsos, G. J., S. D. Collins, and A. D. McClellan (1978). Learning: a model system for physiological studies, *Science (Wash., D.C.)* **199**:497–506.

Mpitsos, G. J. and W. J. Davis (1973). Learning: classical and avoidance conditioning in the mollusk *Pleurobranchaea, Science (Wash., D.C.)* **180**: 317–320.

Muller, K. J. and U. J. McMahan (1976). The shapes of sensory and motor neurones and the distribution of their synapses in ganglia of the leech: a study using intracellular injection of horseradish peroxidase, *Proc. R. Soc. London B Biol. Sci.* **194**:481–499.

Nicholls, J. G. and D. A. Baylor (1968). Specific modalities and receptive fields of sensory neurons in CNS of the leech, *J. Neurophysiol.* **31**:740–756.

Nicholls, J. G. and D. Purves (1970). Monosynaptic chemical and electrical connexions between sensory and motor cells in the central nervous system of the leech, *J. Physiol. (London)* **209**:647–667.

Noble, D. (1977). The initiation of the heart beat, *Proc. Int. Union Physiol. Sci.* **12**:80.

Otsuka, M. and S. Konishi (1976). Substance P and excitatory transmitter of primary sensory neurons, *Cold Spring Harbor Symp. Quant. Biol.* **60**:135–143.

Otsuka, M., E. A. Kravitz, and D. D. Potter (1967). Physiological and chemical architecture of a lobster ganglion with particular reference to gamma-aminobutyrate and glutamate, *J. Neurophysiol.* **30**:725–752.

Pinsker, H. M., W. A. Hening, T. J. Carew, and E. R. Kandel (1973). Long-term sensitization of a defensive withdrawal reflex in *Aplysia, Science (Wash., D.C.)* **182**:1039–1042.

Pinsker, H., I. Kupfermann, V. Castellucci, and E. R. Kandel (1970). Habituation and dishabituation of the gill-withdrawal reflex in *Aplysia, Science (Wash., D.C.)* **167**:1740–1742.

Quinn, W. G., W. A. Harris, and S. Benzer (1974). Conditioned behavior in *Drosophila melanogaster, Proc. Natl. Acad. Sci. U.S.A.* **71**:708–712.

Ramón y Cajal. S. (1909). *Histologie du Système Nerveux de L'homme et des Vertébrés*, 2 v. Maloine, Paris. (Republished 1952–1955, *Histologie du Système Nerveux*, Azoulay, L., trans. Instituto Ramón y Cajal, Madrid.)

Rasmussen, H. (1975). Ions as 'second messengers,' pp. 203–212 in *Cell Membranes: Biochemistry, Cell Biology and Pathology*, Weissmann, G. and R. Claiborne, eds. H.P. Publishing Co., New York.

Razran, G. (1971). *Mind in Evolution.* Houghton Mifflin, Boston.

Rodbell, M. (1975). On the mechanism of activation of fat cell adenylate cyclase by guanine nucleotides: an explanation for the biphasic inhibitory and stimulatory effects of the nucleotides and the role of hormones, *J. Biol. Chem.* **250**:5826–5834.

Schleidt, W. M. (1961*a*) Über die Auslösung der Flucht vor Raubvögeln bei Truthühnern, *Naturwissenschaften* **48**:141–142.

Schleidt, W. M. (1961*b*). Reaktionen von Truthühnern auf fliegende Raubvögel und Versuche zur Analyse ihrer AAM's. *Z. Tierpsychol.* **18**:534–560.

Schneirla, T. C. (1929). Learning and orientation in ants. Studied by means of the maze method, *Comp. Psychol. Monogr.* **6**(4):1–143.

Schwartz, J. H., V. F. Castellucci, and E. R. Kandel (1971). Functioning of identified neurons and synapses in the abdominal ganglion of *Aplysia* in the absence of protein synthesis, *J. Neurophysiol.* **34**:939–953.

Shimahara, T. and L. Tauc (1975). Heterosynaptic facilitation in the giant cell of *Aplysia, J. Physiol. (London)* **247**:321–341.

Singer, S. J. and G. L. Nicolson (1972). The fluid mosaic model of the structure of cell membranes, *Science (Wash., D.C.)* **175**:720–731.

Smith, T. G., Jr., J. L. Barker, and H. Gainer (1975). Requirements for bursting pacemaker potential activity in molluscan neurones, *Nature (London)* **253**: 450–452.

Sokolov, E. N. (1963). *Perception and the Conditioned Reflex,* Waydenfeld, S. W., trans. Pergamon Press, New York.

Spencer, W. A. and E. R. Kandel (1961*a*). Electrophysiology of hippocampal neurons. III. Firing level and time constant, *J. Neurophysiol.* **24**:260–271.

Spencer, W. A. and E. R. Kandel (1961*b*). Electrophysiology of hippocampal neurons. IV. Fast prepotentials, *J. Neurophysiol.* **24**:272–285.

Spencer, W. A., R. F. Thompson, and D. R. Neilson, Jr. (1966*a*). Response decrement of the flexion reflex in the acute spinal cat and transient restoration by strong stimuli, *J. Neurophysiol.* **29**:221–239.

Spencer, W. A., R. F. Thompson, and D. R. Neilson, Jr. (1966*b*). Alterations in responsiveness of ascending and reflex pathways activated by iterated cutaneous afferent volleys, *J. Neurophysiol.* **29**:240–252.

Spencer, W. A., R. F. Thompson, and D. R. Neilson, Jr. (1966*c*). Decrement of ventral root electrotonus and intracellularly recorded PSPs produced by iterated cutaneous afferent volleys, *J. Neurophysiol.* **29**:253–274.

Sperry, R. W. (1951). Mechanisms of neural maturation, pp. 236–280 in *Handbook of Experimental Psychology,* Stevens, S. S., ed. John Wiley & Sons, New York.

Sperry, R. W. (1963). Chemoaffinity in the orderly growth of nerve fiber patterns and connections, *Proc. Natl. Acad. Sci. U.S.A.* **50**:703–710.

Stinnakre, J. and L. Tauc (1973). Calcium influx in active *Aplysia* neurones detected by injected aequorin, *Nature New Biol.* **242**:113–115.

Strumwasser, F. (1967). Types of information stored in single neurons, pp. 291–319 in *Invertebrate Nervous Systems,* Wiersma, C. A. G., ed. University of Chicago Press, Chicago.

Stuart, A. E. (1969). Excitatory and inhibitory motoneurons in the CNS of the leech. Ph.D. Dissertation, Yale University, New Haven.

Susswein, A. J., K. R. Weiss, and I. Kupfermann (1978). The effects of food arousal on the latency of biting in *Aplysia, J. Comp. Physiol. A* **123**:31–41.

Thompson, E. B., C. H. Bailey, V. Castellucci, and E. R. Kandel (1978). A fine structural study of identified sensory and motor neurons which mediate the gill-withdrawal reflex in *Aplysia californica,* in preparation.

Thompson, S. H. and S. J. Smith (1976). Depolarizing afterpotentials and burst production in molluscan pacemaker neurons, *J. Neurophysiol.* **39**: 153–161.

Thorpe, W. H. (1956). *Learning and Instinct in Animals.* Harvard University Press, Cambridge, Mass.

Tinbergen, N. (1951). *The Study of Instinct.* Clarendon Press, Oxford.

Tsien, R. W. (1973). Adrenaline-like effects of intracellular iontophoresis of cyclic AMP in cardiac Purkinje fibres, *Nature New Biol.* **245**:120–122.

Tsien, R. W. (1977). Cyclic AMP and contractile activity in heart, pp. 363–420 in *Advances in Cyclic Nucleotide Research,* Vol. 8, Greengard, P. and G. A. Robison, eds. Raven Press, New York.

Wachtel, H. and E. R. Kandel (1967). A direct synaptic connection mediating both excitation and inhibition, *Science (Wash., D.C.)* **158**:1206–1208.

Wagner, A. R. (1976). Priming in STM: an information-processing mechanism for self-generated or retrieval-generated depression in performance, pp. 95–128 in *Habituation: Perspectives from Child Development, Animal Behavior, and Neurophysiology,* Tighe, T. J. and R. N. Leaton, eds., Lawrence Erlbaum Associates, Hillsdale, N. J.

Weinreich, D. (1977). Synaptic responses mediated by identified histamine-containing neurones, *Nature (London)* **267**:854–856.

Weiskrantz, L. (1970). A long-term view of short-term memory in psychology, pp. 63–74 in *Short-Term Changes in Neural Activity and Behaviour,* Horn, G. and R. A. Hinde, eds. Cambridge University Press, Cambridge, England.

Weiss, K. R., J. Cohen, and I. Kupfermann (1975). Potentiation of muscle contraction: a possible modulatory function of an identified serotonergic cell in *Aplysia, Brain Res.* **99**:381–386.

Weiss, K. R., J. L. Cohen, and I. Kupfermann (1978). Modulatory control of buccal musculature by a serotonergic neuron (metacerebral cell) in *Aplysia, J. Neurophysiol.* **41**:181–203.

Weiss, K. R. and I. Kupfermann (1976). Homology of the giant serotonergic neurons (metacerebral cells) in *Aplysia* and pulmonate molluscs, *Brain Res.* **117**:33–49.

Weiss, K. R. and I. Kupfermann (1978). Serotonergic neuronal activity and arousal of feeding in *Aplysia californica,* pp. 66–89 in *Society for Neuroscience Symposia,* Vol. III, Ferrendelli, J. A., ed. Society for Neuroscience, Bethesda, Md.

Weiss, K. R., M. Schonberg, J. Cohen, D. Mandelbaum, and I. Kupfermann (1976). Modulation of muscle contraction by a serotonergic neuron: possible role of cyclic AMP, *Soc. Neurosci. Abstr.,* Vol. 2, p. 338.

Weiss, K. R., M. Schonberg, D. E. Mandelbaum, and I. Kupfermann (1978). Activity of an individual serotonergic neurone in *Aplysia* enhances synthesis of cyclic adenosine monophosphate, *Nature (London)* **272**:727–728.

Wickelgren, W. A. (1973). The long and the short of memory, *Psychol. Bull.* **80**:425–438.

Wiersma, C. A. G. (1952). The neuron soma; neurons of arthropods, *Cold Spring Harbor Symp. Quant. Biol.* **17**:155–163.

Willows, A. O. D. (1967). Behavioral acts elicited by stimulation of single, identifiable brain cells, *Science* (*Wash., D.C.*) **157**:570–574.

Wilson, W. A. and H. Wachtel (1974). Negative resistance characteristic essential for the maintenance of slow oscillations in bursting neurons, *Science (Wash., D.C.)* **186**:932–934.

Woollacott, M. and G. Hoyle (1977). Neural events underlying learning in insects: changes in pacemaker, *Proc. R. Soc. Lond. B Biol. Sci.* **195**:395–415.

Zucker, R. S. (1972*a*). Crayfish escape behavior and central synapses. I. Neural circuit exciting lateral giant fiber, *J. Neurophysiol.* **35**:599–620.

Zucker, R. S. (1972*b*). Crayfish escape behavior and central synapses. II. Physiological mechanisms underlying behavioral habituation, *J. Neurophysiol.* **35**:621–637.

Zucker, R. S., D. Kennedy, and A. I. Selverston (1971). Neuronal circuit mediating escape responses in crayfish, *Science (Wash., D.C.)* **173**:645–650.

AUTHOR INDEX

SUBJECT INDEX

Harold M. Pinsker